SPACE AND GEOMETRY

IN THE LIGHT OF PHYSIOLOGICAL, PSYCHOLOGICAL, AND PHYSICAL INQUIRY

ERNST MACH

Translated by

Thomas J. McCormack

ROUGH DRAFT PRINTING
1906

ISBN 978-1-60386-436-7

CONTENTS.

ON PHYSIOLOGICAL, AS DISTINGUISHED FROM GEOMETRICAL, SPACE

THE SPACE OF VISION.

The sensible space of our immediate perception, which we find ready at hand on awakening to full consciousness, is considerably different from geometrical space. Our geometrical concepts have been reached for the most part by purposeful experience. The space of the Euclidean geometry is everywhere and in all directions constituted alike; it is unbounded and it is infinite in extent. On the other hand, the space of sight, or "visual space," as it has been termed by Johannes Müller and Hering, is found to be neither constituted everywhere and in all directions alike, nor infinite in extent, nor unbounded.[1] The facts relating to the vision of forms, which I have discussed in another place, show that entirely different feelings are associated with "upness" and "downness," as well as with "nearness" and "farness." "Rightness" and "leftness" are like-

[1] These terms are used in Riemann's sense.

wise the expression of different feelings, although in this case the similarity, owing to considerations of physiological symmetry,[1] is greater. The unlikeness of different directions finds its expression in the phenomena of physiological similarity. The apparent augmentation of the stones at the entrance to a tunnel as we rapidly approach it in a railway train, the shrinkage of the same objects on the train's emerging from the tunnel, are exceptionally distinct cases only of the fact of daily experience that objects in visual space cannot be moved about without suffering expansion and contraction,—so that the space of vision resembles in this respect more the space of the metageometricians than it does the space of Euclid.

Even familiar objects *at rest* exhibit the same peculiarities. A long cylindrical glass vessel tipped over the face, a walking-stick laid endwise against one of the eyebrows, appear strikingly conical in shape. The space of our vision is not only bounded, but at times it appears to have even very narrow boundaries. It has been shown by an experiment of Plateau that an after-image no longer suffers appreciable diminution when projected upon a surface the distance of which from the eye exceeds thirty meters. All ingenuous people, who rely on direct perception, like the astronomers of antiquity, see the heavens approximately as a sphere, finite in

---

[1] *Analysis of the Sensations*, 1886. English trans. Chicago, 1897, p. 49 et seq.

extent. In fact, the oblateness of the celestial vault vertically, — a phenomenon with which even Ptolemy was acquainted, and which Euler has discussed in modern times,—is proof that our visual space is of unequal extent even in different directions. Zoth appears to have found a physiological explanation of this fact, closely related to the conjecture of Ptolemy, in that he interprets the phenomenon as due to the *elevation* of the line of sight with respect to the head.[1] The narrow boundaries of space follow, indeed, directly from the possibility of panoramic painting. Finally, let us observe that visual space in its origin is in nowise metrical. The localities, the distances, etc., of visual space differ only in quality, not in quantity. What we term visual measurement is ultimately the upshot of primitive physical and metrical experiences.

The Space of Touch.

Likewise the skin, which is a closed surface of complicated geometrical form, is an agency of spatial perception. Not only do we distinguish the quality of the irritation, but by some sort of a *supplementary* sensation we also distinguish its *locality*. Now this supplementary sensation need only differ from place to place (the difference in-

[1] Zoth's researches have recently been completed by F. Hillebrand, "Theorie der scheinbaren Groesse bei binocularem Sehen" (*Denkschrift der Wiener Akademie*, math.-naturw. Cl. Bd. 72, 1906).

creasing with the distance apart of the spots irritated) for the purely biological needs of the organism to be satisfied. The great discrepancies that the space-sense of the skin presents with metrical space have been investigated by E. H. Weber.[1] The distance apart at which the two points of a pair of dividers are distinctly recognizable, is from fifty to sixty times less on the tip of the tongue than it is on the middle of the back. At different parts the skin shows great divergencies of spatial sensibility. A pair of dividers the points of which enclose the upper and lower lips, appears sensibly to shut when moved horizontally towards the side of the face (Fig. 1). If the points of the dividers be placed on two adjacent finger-tips and thence carried over the fingers, the palm of the hand, and down the forearm, they will appear at the latter point to close completely (Fig. 2). (The real path of the points is dotted in the figure; the apparent, marked by lines.) The *forms* of bodies that touch the skin are indeed distinguished;[2] but the spatial sense of the skin is nevertheless greatly inferior to that of the eye, although the tip of the tongue will recog-

---

[1] "Ueber den Raumsinn und die Empfindungskreise in der Haut und im Auge" (*Berichte der Kg. Sächs. Gesellsch. der Wissenschaften*, math.-naturw. Cl. 1852, p. 85 et seq.).

[2] Care must be taken that the bodies come into intimate contact with the skin. Various objects having been placed in my paralyzed hand, I was unable to recognize some, and the conclusion was formed that the sensibility of the skin had been impaired. But the conclusion was erroneous; for immediately after the examination, I had another person close my hand and I recognized at once all objects put in it.

nize the circular form of the cross-section of a tube 2 mm. in diameter.

The space of the skin is the analogue of a two-dimensional, finite, unbounded and closed Riemannian space. Through the sensations induced by the

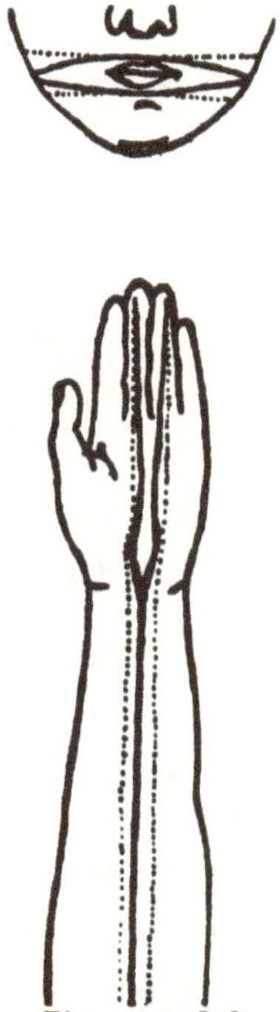

Figs. 1 and 2.

movements of the various members of the body (notably the arms, the hands, and the fingers) something analogous to a third dimension is superposed. Gradually we are led to the interpretation of this system of sensations by the simpler and more salient relations of the physical world. Thus we estimate with considerable exactness the thickness of a plate that we grasp in the dark with the forefinger and thumb of our hand; and we may do the same tolerably well also by touching the upper surface with the finger of one hand and the lower with the finger of the other. Haptic space, or the space of touch, has as little in common with metric space as has the space of vision. Like the latter, it also is anisotropic and non-homogeneous. The cardinal directions of the organism, "forwards and backwards," "upwards and downwards," "right and left," are in both physiological spaces alike non-equivalent.

Sense of Space Dependent on Biological Function.

The fact that our sense of space is not developed at points where it can have no biological function, should not be a cause of special astonishment to us. What purpose could it serve to be informed concerning the location of internal organs over the functions of which we have no control? Thus, our sense of space does not extend to any great distance into the interior of the nostrils. We cannot tell whether we perceive scents introduced by one of a pair of pipettes, at the right or at the left. (E. H. Weber, *loc. cit.*, p. 126.) On the other hand, tactual sensibility, in the case of the ear, according to Weber, extends as far as the tympanum, and enables us to determine whether the louder of two sound-impressions comes from the right or the left. Rough information as to the locality of the source of the sound may be effected in this manner; but it is inadequate for exact purposes.

Correspondence of Physiological and Geometric Space.

Physiological space, thus, has but few qualities in common with geometric space. Both spaces are threefold manifoldnesses. To every point of geometric space, *A, B, C, D,* corresponds a point *A′, B′, C′, D′* of physiological space. If *C* lies between *B* and *D,* then also will *C′* lie between *B′* and *D′*. We may also say that to a continuous motion of a point in geometric space there corresponds a continuous motion of a co-ordinate point in physiolcgical space. I have remarked elsewhere that this continuity, which is merely a convenient fiction, need not in the case of either space be an actual continuity. As every system of sensations, so also the system of space-sensations, is finite,—a fact which cannot astonish us. An endless series of sensational qualities or intensities is psychologically inconceivable. The other properties of visual space also are adapted to biological conditions. The biological needs would not be satisfied with the pure relations of geometric space. "Rightness," "leftness," "aboveness," "belowness," "nearness," and "farness," must be distinguished by a sensational quality. The locality of an object, and not merely its relation to other localities, must be known, if an animal is to profit by such knowledge. It is also advantageous that the sensational indices of visual objects which are near by and consequently more important biologically, are sharply graduated; whereas with the limited stock of indices at hand in the case of remote and less important objects economy is practiced.

A Teleological Explanation.

We shall now develop a simple general consideration, which is again essentially of a teleological nature. Let several distinct spots on the skin of a frog be successively irritated by drops of acid; the frog will respond to each of the several irritations with a specific movement of defense corresponding to the spot irritated. Qualitatively like stimuli affecting different elementary organs and entering by different paths give rise to processes which are propagated back to the environment of the animal again by different organs along different paths. As self-observation shows, we not only recognize the sameness of the irritational quality of a burn at whatever sensitive spot it may occur, but we also distinguish the spots irritated; and our conscious or unconscious movement for protection is executed accordingly. The same holds true for itching, tickling, pressure on the skin, etc. We may be permitted to assume, accordingly, that in all these cases there is resident in the sensation, which qualitatively is the same, some differentiating constituent which is due to the specific character of the elementary organ or spot irritated, or, as Hering would say, to the locality of the attention. Conditions resembling those which hold for the skin doubtless also obtain for the extended surface of any sensory organ; although, as in the case of the retina, the facts are here somewhat more complicated. Instead of movements for protection or flight, may appear also, conformably with the quality of the irritation, movements[1] of attack, the form of which is also determined by the spot irritated. The snapping reflex of the frog, which is produced optically, and the picking of young chicks, may serve as examples. *The perfect biological adaptation of large groups of connected elementary organs among one another is thus very distinctly expressed in the perception of space.*

All Sensation Spatial in Character.

This natural and ingenuous view leads directly to the theory advanced by Prof. William James, according to which *every* sensation is in part spatial in character; a distinct locality, determined by the element irritated, being its invariable accompaniment. Since generally a plurality of elements enters into play, *voluminousness* would also have to

[1] I accept, it will be seen, in a somewhat modified and extended form, the opinion advanced by Wlassak. Cf. his beautiful remarks, "Ueber die statischen Functionen des Ohrlabyrinths," *Vierteljahrssch. f. w. Philos*, XVII. 1 s. 29.

be ascribed to sensations. In support of his hypothesis James frequently refers to Hering. This conception is, in fact, almost universally accepted for optical, tactual, and organic sensations. Many years ago, I myself characterized the relationship of tones of different pitch as spatial, or rather as analogous to spatial; and I believe that the casual remark of Hering, that deep tones occupy a greater volume than high tones, is quite apposite.[1] The highest audible notes of Koenig's rods give as a fact the impression of a needle-thrust, while deep tones appear to fill the entire head. The possibility of localizing sources of sound, although not absolute, also points to a relation between sensations of sound and space. In the first place, we clearly distinguish, in the case of high tones at least, whether the right or the left ear is more strongly affected. And although the parallel between binocular vision and binaural audition, which Steinhauser[2] assumes, may possibly not extend very far, there exists, nevertheless, a certain analogy between them; and the fact remains that the localizing of sources of sound is effected preferentially by the agency of high tones[3]

[1] I am unable to give the reference for this remark definitely; it was therefore doubtless made to me orally. Germs of a similar view, as well as suggestions toward the modern physical theories of audition, are to be found even in Johannes Müller (*Zur vergleich. Physiolog. des Gesichtssinnes*, Leipsic, 1826, p. 455 et seq.).

[2] Steinhauser, *Ueber binaureales Hören*. Vienna. 1877.

[3] "Ueber die Funktion der Ohrmuschel." Tröltsch, *Archiv für Ohrenheilkunde*, N. F., Band 3, S. 72.

(of small volume and more sharply distinguished locality).

Non-Coincidence of the Physiological Spaces.

The physiological spaces of the different senses embrace in general physical domains which are only in part coincident. Almost the entire surface of the skin is accessible to the sense of touch, but only a part of it is visible. On the other hand, the sense of sight, as a telescopic sense, extends in general very much farther physically. We cannot see our internal organs, which, like the elementary organs of sense, we feel as existing in space and invest with locality only when their equilibrium is disturbed; and these same organs fall only partly within range of the sense of touch. Similarly, the determination of position in space by means of the ear is far more uncertain and is restricted to a much more limited field than that by the eye. Yet, loosely connected as the different space-sensations of the different senses may originally have been, they have still entered into connection through association, and that system which has the greater practical importance at the time being is prepared to take the place of the other (James). The space-sensations of the different senses are undoubtedly related, but they are certainly not identical. It is of little consequence whether all these sensations be termed space-sensations or whether *one* species only be invested with this name and the others be conceived as analogues of them.

Sensation in Its Biological Relationship.

If sensation generally, inclusive of sensation of space, be conceived not as an isolated phenomenon, but in its biological functioning, in its biological relationship, the entire subject will be rendered more intelligible. As soon as an organ or system of organs is irritated, the appropriate movements are induced as reflexes. If in complicated biological conditions these movements be found to be evoked spontaneously in response to a part only of the original irritation, in response to some slight impulse, in response to a memory, then we are obliged to assume that traces corresponding to the character of the irritation as well as to that of the irritated organs must be left behind in the memory. It is intelligible thus that every sensory field has its own memory and its own spatial order.

The physiological spaces are multiple manifoldnesses of sensation. The wealth of the manifoldness must correspond to the wealth of the elements irritated. The more nearly elements of the same kind lie together, the more nearly are they akin embryologically, and the more nearly alike are the space-sensations which they produce. If A and B be two elementary organs, it is permissible to assume that the space-sensation produced by each of them is composed of two constituent parts, a and b, of which the one, *a*, diminishes the more, and the other, *b*, increases the more, the farther *B* is removed from *A*, or the more the ontogenetic relationship of *B* to *A* decreases. The elements situated in the series *AB* present a continuously graduated onefold manifoldness of sensation. The multiplicity of the spatial manifoldness must be determined in each case by a special investigation; for the skin, which is a closed surface, a twofold manifoldness would suffice, although a multiple manifoldness is not excluded, and is, by reason of the varying importance of different parts of the skin, even very probable.

It may be said that sensible space consists of a *system of graduated feelings evoked by the sensory organs,* which, while it would not exist without the sense-impressions arising from these organs, yet when aroused by the latter *constitutes a sort of scale in which our sense-impressions are registered.* Although every single feeling due to a sensory organ (feeling of space) is registered according to its specific character between those next related to it, a plurality of excited organs is nevertheless very advantageous for distinctness of localization, for the reason that the contrasts between the feelings of locality are enlivened in this way. Visual space, therefore, which ordinarily is well filled with objects, thus affords the best means of localization. Localization becomes at once uncertain and fluctuant for a single bright spot on a dark background.

Origin of the Three Dimensions.

It may be assumed that the system of space-sensations is in the main very similar, though unequally developed, in all animals which, like man, have three cardinal directions distinctly marked on their bodies. Above and below, the bodies of such animals are unlike, as they are also in front and behind and to the right and to the left. To the right and the left, these animals are apparently alike, but their geometrical and mechanical symmetry, which subserves purposes of rapid locomotion, should not deceive us with regard to their anatomical and physiological asymmetry. Though the latter may appear slight, it is yet distinctly marked in the fact that species very closely allied to symmetrical animals sometimes assume strikingly unsymmetrical forms. The asymmetry of the plaice (flatfish) is a familiar instance, while the externally symmetric form of the slug forms an instructive contrast to the unsymmetric shapes of some of its nearer relatives. This trinity of conspicuously marked cardinal directions might indeed be regarded as the physiological basis for our familiarity with the three dimensions of geometric space.

Biological Importance of Tactual Space.

Visual space forms the clearest, precisest, and broadest system of space-sensations; but, biologically, tactual space is perhaps more important. Irritations of the skin are spatially registered from the very outset; they disengage the corresponding protective movement; the disengaged movement then again induces sensations in the extended or contracted skin, in the joints, in the muscles, etc., which are associated with sensations of space. The first localizations in tactual space are presumably effected on the body itself; as when the palm of the hand, for example, is carried over the surface of the thigh, which also is sensitive to impressions of space. In this manner are experiences in the field of tactual space gathered. But the attempt which is frequently made of deriving tactual space psychologically from such experiences, by aid of the concept of time and on the assumption of spaceless sensations, is an altogether futile one.

Visual and Tactual Space Correlated.

It is my opinion that the space of touch and the space of vision may be conceived after quite the same manner. This can be done (so far as I can infer from what has already been attempted in this direction) only by transferring Hering's view of visual space to tactual space. This also accords best with general biological considerations. A newly-hatched chick notices a small object, looks toward it, and immediately pecks at it. A certain area in the central organ is excited by the irritation, and the looking movement of the muscles of the eye, as well as the picking movements of the head and neck, are forthwith automatically disengaged thereby. The excitation of the above-mentioned area of the central organ, which on the one hand is determined by the geometric locality of the physical irritation, is on the other hand the basis of the space-sensation. The disengaged muscular movements themselves become a source of sensations in greatly varying degree. Whereas the sensations attending the movements of the eyes, in the case of man at least, usually disappear almost altogether, the movements of the muscles made in the performance of work leave behind them a powerful impression. The behavior of the chick is quite similar to that of an infant which spies a shining object and snatches at it.

It will scarcely be questioned that in addition to optical irritations other irritations, acoustic, thermal, and gustatory in character, are also able to evoke movements of prehension or defense, especially so in the case of blind people, and that to the same movements, the same irritated parts of the central organ, and therefore also the same sensation of space, will correspond. The irritations affecting blind people are, as a general thing, merely limited to a more restricted sphere and less sharply determined as to locality. The system of spatial sensations of such people must at first be rather meager and obscure; consider, for instance, the situation of a blind person endeavoring to protect himself from a wasp buzzing around his head. Yet education can do very much towards perfecting the spatial sense of blind people, as the achievements of the blind geometer Saunderson clearly show. Spatial orientation must notwithstanding have been somewhat difficult for him, as is proved by the construction of his table, which was divided in the simplest manner into quadratic spaces. He was wont to insert pins into the corners and centers of these squares and to connect their heads by threads. His highly original work, however, must by reason of its very simplicity have been particularly easy for beginners to understand; thus he demonstrated the proposition that the volume of a pyramid is equal to one-third of the volume of a prism of the same base and height by dividing a cube into six congruent pyramids, each having a side of the cube for its base and its vertex in the center of the cube.[1]

Tactual space exhibits the same peculiarities of anisotropy and of dissimilarity in the three cardinal directions as visual space, and differs in these peculiarities also from the geometric space of Euclid. On the other hand, optical and tactual space-sensations are at many points in accord. If I stroke with my hand a stationary surface having upon it distinct tangible objects, I shall feel these objects as at rest, just as I should feel visual objects to be when voluntarily causing my eyes to pass over them, although the images themselves actually move across the retina. On the other hand, a moving object

[1] Diderot, *Lettre sur les aveugles.*

appears in motion to the seeing or touching organ either when the latter is at rest or when it is following the object. Physiological symmetry and similarity find the same expression in the two domains, as has been elsewhere shown in detail;[1] but, however intimately allied they may be, the two systems of space-sensations cannot nevertheless be identical. When an object excites me in one case to look at it and in another to grasp it, certainly the portions of the central organ which are affected must be in part different, no matter how nearly contiguous they may be. If both results take place, the domain is naturally larger. For biological reasons, we may expect that the two systems readily coalesce by association, and readily adapt themselves to one another, as is actually the case.

Feelings of Space Involve Stimulus to Motion.

But the province of the phenomena with which we are concerned is not yet exhausted. A chick can look at an object, pick at it, or even be determined by the stimulus presented to run to it, turn towards or around to it. A child that is creeping toward an objective point, and then some day gets up and runs with several steps toward it, acts likewise. We are under the necessity of conceiving these cases, which pass continuously into one another, from some similar point of view. There must be certain parts of

[1] *Analysis of the Sensations*, Eng. trans., p. 50 et seq.

the brain which, having been irritated in a comparatively simple manner, on the one hand give rise to feelings of space and on the other hand, by their organization, produce automatic movements which at times may be quite complicated. The stimulus to extensive locomotion and change of orientation not only proceeds from optical excitations, but may also be induced, even in the case of blind animals, by chemical, thermal, acoustic, and galvanic excitations.[1] In point of fact, we also observe extensive movements of locomotion and orientation in animals that are constitutionally blind (blind worms), as well as in such as are blind by retrogression (moles and cave animals). We may accordingly conceive sensations of space as determined in a perfectly analogous manner both in animals with and in animals without sight.

A person watching a centipede creeping uniformly along is irresistibly impressed with the idea that there proceeds from some organ of the animal a uniform stream of stimulation which is answered by the motor organs of its successive segments with rhythmic automatic movements. Owing to the difference of phase of the hind as compared with the fore segments, there is produced a longitudinal wave which we see propagated through the legs of the animal with mechanical regularity. Analogous phenomena cannot be wanting in the higher ani-

[1] Loeb, *Vergleichende Gehirnphysiologie*, Leipzig, 1899, page 108 et seq.

mals, and as a matter of fact do exist there. We have an analogous case during active or passive rotation about the vertical axis, when the irritation induced in the labyrinth disengages the well known nystagmic movements of the eyes. The organism adapts itself so perfectly to certain regular alterations of excitations that on the cessation of these alterations under certain circumstances negative after-images are produced. I have but to recall to the reader's mind the experiment of Plateau and Oppel with the expanding spiral, which when brought to rest appears to shrink, and the corresponding results which Dvorák produced by alterations of the intensity of light. Phenomena of this kind led me long ago to the assumption that there corresponded to an alteration of the stimulus u with the time t, to a rate of alteration, $\frac{du}{dt}$ a special process which under certain circumstances might be felt and which is of course associated with some definite organ. Thus, rate of motion, within the limits within which the perceiving organ can adapt itself, is felt directly; this is therefore not only an abstract idea, as is the speed of the hand of a clock or of a projectile, but it is also a specific sensation, and furnished the original impulse to the formation of the idea. Thus, a person feels in the case of a line not only a succession of points varying in position, but also the direction and the curvature of the line. If the intensity of illumination of a surface is given by $u =$

$f(x, y)$, then not only u but also $\frac{du}{dx}$, $\frac{du}{dy}$ and $\frac{d^2u}{dx^2}$, $\frac{d^2u}{dy^2}$ find their expression in sensation,—a circumstance which points to a complicated relationship between the elementary organs.

The Central Motor Organ and the Will to Move.

If there actually exists, then, as in the centipede, an organ which on simple irritation disengages the complicated movements belonging to a definite kind of locomotion, it will be permissible to regard this simple irritation, provided it is conscious, as *the will* or *the attention* appurtenant to this locomotion and carrying the latter spontaneously with it. At the same time, it will be recognized as a need of the organism that the effect of the locomotion should be felt in a correspondingly simple manner.

Biological Necessity Paramount.

For detailed illustration, we will revert once more to the consideration of visual space. The perception of space proceeds from a biological need, and will be best understood in its various details from this point of view. The greater distinctness and the greater nicety of discrimination exercised at a single specific spot on the retina of vertebrate animals is an economic device. By it, the possibility of moving the eye in response to changes of attention is rendered necessary, but at the same time the disturbing effects of *willed* movements of the eyes on the sensations of space induced by objects at rest have to be excluded. Perception of the movement of an image across the retina when the retina is at rest, perception of the movement of an object when the eye is at rest, is a biological necessity. As for the perception of objects at rest in the unfrequent contingency of a movement of the eye due to some occurrence extrinsic to consciousness (external mechanical pressure, or twitching of the muscles), this was unnecessary for the organism. The foregoing requirements are to be harmonized only on the assumption that the displacement of the image on the retina of the eye in voluntary movement is offset as to spatial value by the volitional character of the movement. It follows from this that objects at rest may be made, while the eye also is at rest, to suffer displacement in visual space by the *tendency to movement* merely, as has been actually shown by experiment.[1] The second offsetting factor is also directly indicated in this experiment. The organism is not obliged, further, in accomplishing its adaptation, to take account of the second contingency mentioned, which arises only under pathological or artificial circumstances. Paradoxical as the conditions here involved may appear, and far removed as we may still be from a *causal* comprehension of them, they are nevertheless easily under-

[1] *Analysis of the Sensations.* English Trans. Page 59.

stood when thus viewed teleologically as a connected whole.

Sensations of Movement.

Shut up in a cylindrical cabinet rotating about a vertical axis, we see and feel ourselves rotating, along with the cylindrical wall, in the direction in which the motion takes place. The impression made by this sensation is at first blush highly paradoxical, inasmuch as there exists not a vestige of a reason for our supposing that the rotation is a relative one. It appears as if it would be actually possible for us to have sensations of movement in absolute space,—a conception to which no physical significance can possibly be attached. But physiologically the case easily admits of explanation. An excitation is produced in the labyrinthine canals of the internal ear,[1] and this excitation disengages, independently of consciousness, a reflex rotary movement of the eyes in a direction opposite to that of the motion,[2] by which the retinal images of all objects resting against the body are displaced exactly as if they were rotating in the direction of the motion. Fixing the eyes intentionally upon some such object, the rotation does not, as might be supposed, disappear. The eye's tendency to motion is then exactly counterbalanced by the introduction of a fac-

[1] *Bewegungsempfindungen*, 41 et seq. Leipsic, 1875.

[2] Breuer, *Vorläufige Mittheilung im Anzeiger der kk. Gesellschaft der Aerzte in Wien, vom 20. Nov. 1873.*

tor extrinsic to consciousness.[1] We have here the case mentioned above, where the eye, held externally at rest, becomes aware of a displacement in the direction of its tendency to motion. But what before appeared as a paradoxical exception is now a natural result of the adaptation of the organism, by which the animal perceives the motion of its own body when external objects at rest remain stationary. Analogous adaptive results with which even Purkynje was in part acquainted are met with in the domain of the tactile sense.[2]

The eyes of an observer watching the water rushing underneath a bridge are impelled without noticeable effort to follow the motion of the flowing water and to adapt themselves to the same. If the observer will now look at the bridge, he will see both the latter and himself moving in a direction opposite to that of the water. Here again the eye which fixates the bridge must be maintained at rest by a willed motional effort made in opposition to its unconsciously acquired motional tendency, and it now sees apparent motions to which no real motions correspond.

But the same phenomena which appear here paradoxical and singular undoubtedly serve an important function in the case of progressive motion or

---

[1] *Analysis of the Sensations.* English Trans. Page 71.

[2] Purkynje, "Beiträge zur Kenntniss des Schwindels." *Medizin. Jahrbücher des österreichischen Staates,* VI. Wien, 1820. "Versuche über den Schwindel, *10th Bulletin der naturw. Section der schles. Gesellschaft.* Breslau, 1825, s. 25.

locomotion. To the property of the visual apparatus referred to is due the fact that an animal in progressive motion sees itself moving and the stationary objects in its environment at rest.[1] Anomalies of this character, where a body appears to be in motion without moving from the spot which it occupies, where a body contracts without really growing smaller (which we are in the habit of calling illusions on the few rare occasions when we notice them) have accordingly their important normal and common function.

As the process which we term the *will* to turn round or move forward is of a very simple nature, so also is the result of this will characterized by feelings of a very simple nature. *Fluent* spatial values of certain objects, instead of stable, make their appearance in the domain of the tactual as well as the visual sense. But even where visual and tactual sensations are as much as possible excluded, unmistakable sensations of motion are produced; for example, a person placed in a darkened room, with closed eyes, on a seat affording support to the body on all sides, will be conscious of the slightest progressive or angular acceleration in the movement of his body, no matter how noiselessly and gently the same may be produced.[2] By association, these simple sensations also are translated at once into the motor images of the other senses. Between

[1] *Analysis of the Sensations.* English Trans. Pages 63, 64, 71, 72.

[2] *Bewegungsempfindungen*, Leipsic, 1875.

this initial and terminal link of the process are situated the various sensations of the extremities moved, which ordinarily enter consciousness, however, only when obstructions intervene.

Primary and Secondary Space.

We have now, as I believe, gained a fair insight into the nature of sensations of space. The last-discussed species of sensations of space, which were denominated sensations of movement, are sharply distinguished from those previously investigated, by their *uniformity* and *inexhaustibility.* These sensations of movement make their appearance only in animals that are free to move about, whereas animals that are confined to a single spot are restricted to the sensations of space first considered, which we shall designate *primary* sensations of space, as distinguished from *secondary* sensations (of movement). A *fixed* animal possesses necessarily a *bounded* space. Whether that space be symmetrical or unsymmetrical depends upon the conditions of symmetry of its own body. A vertebrate animal confined to a single spot and restricted as to orientation could construct only a bounded space which would be dissimilar above and below, before and behind, and accurately speaking also to the right and to the left, and which consequently would present a sort of analogy with the physical properties of a triclinic crystal. If the animal acquired the power of moving freely about, it would obtain in this way in addition an *infinite* physiological space; for the sensations of movement always admit of being *produced anew* when not prevented by accidental external hindrances. Untrammeled orientation, the interchangeability of every orientation with every other, invests physiological space with the property of equality in all directions. Progressive motion and the possibility of orientation in any direction together render space identically constituted at all places and in all directions. Nevertheless, we may remark at this juncture that the foregoing result has not been obtained through the operation of physiological factors exclusively, for the reason that orientation with respect to the vertical, or the direction of the acceleration of gravity, is not altogether optional in the case of any animal. Marked disturbances of orientation with respect to the vertical make themselves most strongly felt in the higher vertebrate animals by their physico-physiological results, by which they are restricted as regards both duration and magnitude. Primary space cannot be absolutely supplanted by secondary space, for the reason that it is phylogenetically and ontogenetically older and stronger. If primary space decreases in significance during motion, the sensation of movement in its turn immediately vanishes when the motion ceases, as does every sensation which is not kept alive by reviviscence and contrast. Primary space then again enters upon its rights. It is doubtless unnecessary to remark that physiological space is in no wise concerned with metrical relations.

Biological Theory of Spatial Perception.

We have assumed that *physiological* space is an adaptive result of the interaction of the elementary organs, which are constrained to live together and are thus absolutely dependent upon co-operation, without which they would not exist. Of cardinal and greatest importance to animals are the parts of *their own body* and their relations to one another; outward bodies come into consideration only in so far as they stand in some way in relation to the parts of the animal body. The conditions here involved are physiological in character,—which does not exclude the fact that every part of the body continues to be a part of the physical world, and so subject to general physical laws, as is most strikingly shown by the phenomena which take place in the labyrinth during locomotion, or by a change of orientation. *Geometric* space embraces only the relations of physical bodies *to one another*, and leaves the animal body in this connection altogether out of account.

We are aware of but one species of elements of consciousness: sensations. In our perceptions of space we are dependent on sensations. The character of these sensations, and the organs that are in operation while they are felt, are matters that must be left undecided.

The view on which the preceding reflections are based is as follows: The feeling with which an elementary organ is affected when in action, depends partly upon the character (or quality) of the irritation; we will call this part the *sense-impression.* A second part of the feeling, on the other hand, may be conceived as determined by the *individuality* of the organ, being the same for every stimulus and varying only from organ to organ, the degree of variation being inversely proportional to the ontogenetic relationship. This portion of the feeling may be called the *space-sensation.* Space-sensation can accordingly be produced only when there is some irritation of elementary organs; and every time the same organ or the same complexus of organs is irritated, every time the same concatenation of organs is aroused, the same space-sensation is evoked. We make only the same assumptions here with regard to the elementary organs that we should deem ourselves quite justified in making with respect to isolated individual animals of the same phylogenetic descent but different degrees of affinity.

The prospect is here opened of a phylogenetic and ontogenetic understanding of spatial perception; and after the conditions of the case have been once thoroughly elucidated, a physical and physiological explanation seems possible. I am far from thinking that the explanation here offered is absolutely adequate or exhaustive on all sides; but I am convinced that I have made some approach to the truth by it.

The A Priori Theory of Space.

Kant asserted that "one could never picture to oneself that space did not actually exist, although one might quite easily imagine that there were no objects in space." To-day, scarcely any one doubts that sensations of objects and sensations of space can enter consciousness only *in combination* with one another; and that, *vice versa,* they can leave consciousness only in combination with one another. And the same must hold true with regard to the concepts which correspond to these sensations. If for Kant space is not a "concept," but a "pure (mere?) intuition *a priori,*" modern inquirers on the other hand are inclined to regard space as a concept, and in addition as a concept which has been derived from experience. We cannot intuite our system of space-sensations *per se:* but we may neglect sensations of objects as something subsidiary; and if we overlook what we have done, the notion may easily arise that we are actually concerned with a pure intuition. If our sensations of space are independent of the quality of the stimuli which go to produce them, then we may make predications concerning the former independently of external or physical experience. It is the imperishable merit of Kant to have called attention to this point. But this basis is unquestionably inadequate to the complete development of a geometry, inasmuch as concepts, and in addition thereto concepts derived from experience, are also requisite to this purpose.

Physiological Influences in Geometry.

Physiological, and particularly visual, space appears as a distortion of geometrical space when derived from the metrical data of geometrical space. But the properties of continuity and threefold manifoldness are preserved in such a transformation, and all the consequences of these properties may be derived without recourse to physical experience, by our representative powers solely.

Since physiological space, as a system of sensations, is much nearer at hand than the geometric concepts that are based thereon, the properties of physiological space will be found to assert themselves quite frequently in our dealings with geometric space. We distinguish near and remote points in our figures, those at the right from those at the left, those at the top from those at the bottom, entirely by physiological considerations and despite the fact that geometric space is not cognizant of any relation to our body, but only of relations of the points to one another. Among geometric figures, the straight line and the plane are especially marked out by their physiological properties; as they are indeed the first objects of geometrical investigation. Symmetry is also distinctly revealed by its physiological properties, and attracts thus immediately the attention of the geometer. It has doubtless also been efficacious in determining the division of space into right angles. The fact that similitude was investigated previously to other geometric affinities likewise is due to physiological facts. The Cartesian geometry of co-ordinates in a manner liberated geometry from physiological influences, yet vestiges of their thrall still remain in the distinction of positive and negative co-ordinates, according as these are reckoned to the right or to the left, upward or downward, and so on. This is convenient, but not necessary. A fourth co-ordinate plane, or the determination of a point by its distances from four fundamental points not lying in the same plane, exempts geometric space from the necessity of constantly recurring to physiological space. The necessity of such restrictions as "around to the right" and "around to the left," and the distinction of symmetrical figures by these means would then be eliminated. The historical influences of physiological space on the development of the concepts of geometric space are, of course, not to be eliminated.

Also in other provinces, as in physics, the influence of the properties of physiological space is traceable, and not alone in geometry. Even secondary physiological space is considerably different from Euclidean space, owing to the fact that the distinction between "above" and "below" does not absolutely disappear in the former. Sosikles of Corinth (Herodotus v. 92) asseverated that "sooner should the heavens be beneath the earth and the earth soar in the air above the heavens, than that the Spartans should lose their freedom." And his assertion, together with the tirades of Lactantius

(*De falsa sapientia*, c. 24) and St. Augustine (*De civitate dei*, XVII., 9), against the doctrine of the antipodes, against men hanging with inverted heads and trees growing downward,—considerations which even after centuries touch in us a sympathetic chord,—all had their good physiological grounds. We have, in fact less reason to be astonished at the narrow-mindedness of these opponents of the doctrine of the antipodes than we have to be filled with admiration for the great powers of abstraction exhibited by Archytas of Tarentum and Aristarchus of Samos.

ON THE PSYCHOLOGY AND NATURAL DEVELOPMENT OF GEOMETRY.

For the animal organism, the relations of the different parts of *its own body* to one another, and of physical objects to these different parts, are *primarily* of the greatest importance. Upon these relations is based its system of physiological sensations of space. More complicated conditions of life, in which the simple and direct satisfaction of needs is impossible, result in an augmentation of intelligence. The physical, and particularly the *spatial,* behavior of bodies *toward one another* may then acquire a mediate and indirect interest far transcending our interest in our momentary sensations. In this way, a spatial image of the world is created, at first instinctively, then in the practical arts, and finally scientifically, in the form of geometry. The mutual relations of bodies are geometrical in so far as they are determined by sensations of space, or find their expression in such sensations. Just as without sensations of heat there would have been no theory of heat, so also without sensations of space there would be no geometry; but both the theory of heat and the theory of geometry stand additionally in need of *experiences concerning bodies;* that is to say, both must pursue *their inquiries* beyond the narrow boundaries of the domains of sense that constitute their peculiar foundation.

The Rôle of Bodies.

Isolated sensations have *independent* significance only in the lowest stages of animal life; as, for example, in reflex motions, in the removal of some disagreeable irritation of the skin, in the *snapping* reflex of the frog, etc. In the higher stages, attention is directed, not to space-sensation alone, but to those intricate and intimate *complexes* of other sensations with space-sensations which we call *bodies*. Bodies arouse our interest; they are the *objects* of our activities. But the *character* of our activities is coincidently determined by the *place* of the body, whether near or far, whether above or below, etc.,—in other words, by the space-sensations characterizing that body. The *mode* of reaction is thus determined by which the body can be reached, whether by extending the arms, by taking few or many steps, by hurling missiles, or what not. The *quantity* of sensitive elements which a body excites, the number of places which it covers, that is to say, the *volume* of the body, is, all other things being the same, proportional to its capacity for satisfying our needs, and possesses a consequent biological import. Although our sensations of sight and touch are primarily produced only by the *surfaces* of bodies, nevertheless powerful associations impel especially primitive man to imagine more, or, as he thinks, to *per-*

ceive more, than he actually observes. He imagines to be filled with *matter* the places enclosed by the surface which alone he perceives; and this is especially the case when he sees or seizes bodies with which he is in some measure familiar. It requires considerable power of abstraction to bring to consciousness the fact that we perceive the surfaces *only* of bodies,—a power which cannot be ascribed to primitive man.

Of importance in this regard are also the peculiar *distinctive shapes* of objects of prey and utility. Certain definite forms, that is, certain specific combinations of space-sensations, which man learns to know through intercourse with his environment, are unequivocally characterized even by purely physiological features. The straight line and the plane are distinguished from all other forms by their physiological simplicity, as are likewise the circle and the sphere. The affinity of symmetric and geometrically similar forms is revealed by purely physiological properties. The variety of shapes with which we are acquainted from our physiological experience is far from being inconsiderable. Finally, through employment with bodily objects, *physical* experience also contributes its quota of wealth to the general store.

The Notion of Constancy.

Crude physical experience impels us to attribute to bodies a certain *constancy*. Unless there are special reasons for not doing so, the same constancy is also ascribed to the individual attributes of the complexus "body";[1] thus we also regard the color, hardness, shape, etc., of the body as constant; and particularly we look upon the body as *constant with respect to space, as indestructible.* This assumption of spatial constancy, of *spatial substantiality,* finds its direct expression in geometry. Our physiological and psychological organization is independently predisposed to emphasize constancy; for general physical constancies must necessarily have found lodgment in our organization, which is itself physical, while in the adaptation of the species very definite physical constancies were at work. Inasmuch as memory revives the images of bodies, before perceived, in their original forms and dimensions, it supplies the condition for the recognition of the same bodies, thus furnishing the first foundation for the impression of constancy. But geometry is additionally in need of certain *individual* experiences.

Let a body *K* move away from an observer *A* by being suddenly transported from the environment *FGH* to the environment *MNO.* To the optical observer *A* the body *K* decreases in size and assumes generally a different form. But to an optical observer *B,* who moves along with *K* and who always retains the same position with respect to *K, K* remains unaltered. An analogous sensation is experienced by the *tactual observer,* although the per-

[1] See my *Analysis of the Sensations,* introductory chapter.

spective diminution is here wanting for the reason that the sense of touch is not a telepathic sense. The experiences of *A* and *B* must now be harmonized and their contradictions eliminated,—a requirement which becomes especially imperative when *the same* observer plays alternately the parts of *A* and of *B*. And the only method by which they can be harmonized is, to attribute to *K* certain *constant* spatial properties independently of its position with respect to *other* bodies. The space-sensations determined by *K* in the observer *A* are recognized as *dependent* on other space-sensations (the position of *K* with respect to the body of the observer *A*). But these same space-sensations determined by *K* in *A* are *independent* of other space-sensations, characterizing the position of *K* with respect to *B*, or with respect to *FGH...MNO*. In *this* independence lies the *constancy* with which we are here concerned.

The fundamental assumption of geometry thus reposes on an *experience*, although on one of an idealized kind.

The Notion of Rigidity.

In order that the experience in question may assume palpable and perfectly determinate form, the body *K* must be a so-called *rigid* body. If the space-sensations associated with *three* distinct acts of sense-perception remain unaltered, then the condition is given for the invariability of the entire complexus of space-sensations determined by a rigid body. This determination of the space-sensations produced by a body by means of *three* space-sensational *elements* accordingly characterizes the rigid body, from the point of view of the physiology of the senses. And this holds good for both the visual and the tactual sense. In so doing we are not thinking of the physical conditions of rigidity (in defining which we should be compelled to enter different sensory domains), but merely of the fact given to our spatial sense. Indeed, we are now regarding every body as rigid which possesses the property assigned, even liquids, so long as their parts are not in motion with respect to one another.

PHYSICAL ORIGIN OF GEOMETRY.

Correct as the oft-repeated asseveration is that geometry is concerned, not with *physical,* but with *ideal* objects, it nevertheless cannot be doubted that geometry has sprung from the interest centering in the spatial relations of *physical bodies.* It bears the distinctest marks of this origin, and the course of its development is fully intelligible only on a consideration of this fact. Our knowledge of the spatial behavior of bodies is based upon a *comparison* of the space-sensations produced by them. Without the least artificial or scientific assistance we acquire abundant experience of space. We can judge approximately whether rigid bodies which we perceive alongside one another in different positions at different distances, will, when brought *successively* into the same position, produce approximately the same or dissimilar space-sensations. We know fairly well whether one body will coincide with another,—whether a pole lying flat on the ground will reach to a certain height. Our sensations of space are, however, subject to physiological circumstances, which can never be absolutely identical for the members compared. In every case, rigorously viewed, a memory-trace of a sensation is necessarily compared with a real sensation. If, therefore, it is a question of the exact spatial relationship of bodies *to one another,* we must provide characteristics that depend as little as possible on physiological conditions, which are so difficult to control.

Measurement.

This is accomplished by comparing *bodies* with *bodies.* Whether a body *A* coincides with another body *B*, whether it can be made to occupy exactly the space filled by the other—that is, whether under like circumstances both bodies produce the same space-sensations—can be estimated with great precision. We regard such bodies as spatially or geometrically equal in every respect,—*as congruent.* The *character* of the sensations is here no longer authoritative; it is now solely a question of their *equality* or *inequality.* If both bodies are rigid bodies, we can apply to the second body *B* all the experiences which we have gathered in connection with the first, more convenient, and more easily transportable, standard body *A*. We shall revert later to the circumstance that it is neither necessary nor possible to employ a special body of comparison, or standard, for every body. The most convenient bodies of comparison, though applicable only after a crude fashion,—bodies whose invariance during transportation we always have before our eyes,—are our *hands* and *feet*, our *arms* and *legs*. The names of the oldest measures show distinctly that originally we made our measurements with hands'-breadths, forearms (*ells*), feet (*paces*), etc. Nothing but a period of *greater exactitude* in measurement began with the introduction of conventional and carefully preserved physical standards; the principle remains the same. The measure enables us to compare bodies which are difficult to move or are practically immovable.

The Rôle of Volume.

As has been remarked, it is not the *spatial*, but predominantly the *material*, properties of bodies that possess the strongest interest. This fact certainly finds expression even in the beginnings of geometry. The *volume* of a body is instinctively taken into account as representing the quantity of its material properties, and so comes to form an object of *contention* long before its geometric properties receive anything approaching to profound consideration. It is here, however, that the comparison, the measurement of volumes acquires its initial import, and thus takes its place among the first and most important problems of primitive geometry.

The first measurements of volume were doubtless of liquids and fruit, and were made with hollow measures. The object was to ascertain conveniently the quantity of like matter, or the *quantity* (*number*) of homogeneous, similarly shaped (identical) *bodies*. Thus, conversely, the capacity of a store-room (granary) was in all likelihood originally estimated by the quantity or number of homogeneous bodies which it was capable of containing. The measurement of volume by a unit of volume is in all probability a much later conception, and can only have developed on a higher stage of abstraction. Estimates of areas were also doubtless made from the *number* of fruit-bearing or useful plants which a field would accommodate, or from the quantity of seed that could be sown on it; or possibly also from the *labor* which such work required.

Measurement of Surfaces.

The measurement of a surface by a surface was readily and obviously suggested in this connection when fields of the same size and shape lay near one another. There one could scarcely doubt that the field made up of n fields of the same size and form possessed also n-fold agricultural value. We shall not be inclined to underrate the significance of this intellectual step when we consider the errors in the measurement of areas which the Egyptians[1] and even the Roman *agrimensores*[2] commonly committed.

[1] Eisenlohr, *Ein mathematisches Handbuch der alten Aegypter: Papyrus Rhind*, Leipsic, 1877.

[2] M. Cantor, *Die römischen Agrimensoren*, Leipsic, 1875.

Even with a people so splendidly endowed with geometrical talent as the Greeks, and in so late a period, we meet with the sporadic expression of the idea that surfaces having equal perimeters are equal in area.[1] When the Persian "Overman," Xerxes,[2] wished to count the army which was his to destroy, and which he drove under the lash across the Hellespont against the Greeks, he adopted the following procedure. Ten thousand men were drawn up closely packed together. The area which they covered was surrounded with an enclosure, and each successive division of the army, or rather, each successive herd of slaves, that was driven into and filled the pen, counted for another ten thousand. We meet here with the converse application of the idea by which a surface is measured by the *quantity (number) of equal, identical, immediately adjacent bodies which cover it.* In abstracting, first instinctively and then consciously, from the height of these bodies, the transition is made to measuring surfaces by means of a unit of surface. The analogous step to measuring volumes by volume demands a far more practiced, geometrically disciplined intuition. It is effected later, and is even at this day less easy to the masses.

All Measurement by Bodies.

The oldest estimates of long *distances,* which were computed by days' journeys, hours of travel, etc.,

[1] Thucydides, VI., 1.

[2] Herodotus, VII., 22, 56, 103, 223.

were based doubtless upon the effort, labor, and expenditure of time necessary for covering these distances. But when lengths are measured by the repeated application of the hand, the foot, the arm, the rod, or the chain, then, accurately viewed, the measurement is made by the enumeration of like bodies, and we have again really a measurement by volume. The singularity of this conception will disappear in the course of this exposition. If, now, we abstract, first instinctively and then consciously, from the two transverse dimensions of the bodies employed in the enumeration, we reach the measuring of a line by a line.

A surface is commonly defined as the boundary of a space. Thus, the surface of a metal sphere is the boundary between the metal and the air; it is not part either of the metal or of the air; two dimensions only are ascribed to it. Analogously, the one-dimensional line is the boundary of a surface; for example, the equator is the boundary of the surface of a hemisphere. The dimensionless point is the boundary of a line; for example, of the arc of a circle. A point, by its motion, generates a one-dimensional line, a line a two-dimensional surface, and a surface a three-dimensional solid space. No difficulties are presented by this concept to minds at all skilled in abstraction. It suffers, however, from the drawback that it does not exhibit, but on the contrary artificially conceals, the natural and actual way in which the abstractions have been reached. A certain discomfort is therefore felt when the attempt is made from this point of view to define the measure of surface or unit of area after the measurement of lengths has been discussed.[1]

A more homogeneous conception is reached if *every* measurement be regarded as a counting of space by means of immediately *adjacent*, spatially *identical*, or at least hypothetically identical, *bodies*, whether we be concerned with volumes, with surfaces, or with lines. Surfaces may be regarded as corporeal sheets, having everywhere the same constant thickness which we may make small at will, *vanishingly* small; lines, as strings or threads of constant, vanishingly small thickness. A point then becomes a small corporeal space from the extension of which we purposely abstract, whether it be part of another space, of a surface, or of a line. The bodies employed in the enumeration may be of any smallness or any form which conforms to our needs. Nothing prevents our idealizing in the usual manner these images, reached in the natural way indicated, by simply leaving out of account the thickness of the sheets and the threads.

The usual and somewhat timid mode of presenting the fundamental notions of geometry is doubtless due to the fact that the infinitesimal method which freed mathematics from the historical and accidental shackles of its early elementary form, did

[1] Hölder, *Anschauung und Denken in der Geometrie*, Leipsic, 1900, p. 18. W. Killing, *Einführung in die Grundlagen der Geometrie*, Paderborn, 1898, II., p. 22 et seq.

not begin to influence geometry until a later period of development, and that the frank and natural alliance of geometry with the *physical* sciences was not restored until still later, through Gauss. But why the elements shall not now partake of the advantages of our better insight, is not to be clearly seen. Even Leibnitz adverted to the fact that it would be more rational to begin with the *solid* in our geometrical definitions.[1]

Method of Indivisibles.

The measurement of spaces, surfaces, and lines by means of *solids* is a conception from which our refined geometrical methods have become entirely estranged. Yet this idea is not merely the forerunner of the present idealized methods, but it plays an important part in the psychology of geometry, and we find it still powerfully active at a late period of development in the workshop of the investigators and inventors in this domain.

Cavalieri's Method of Indivisibles appears best comprehensible through this idea. Taking his own illustration, let us consider the surfaces to be compared (the quadratures) as covered with equidistant parallel threads of any number we will, after the manner of the warp of woven fabrics, and the spaces to be compared (the cubatures) as filled with parallel sheets of paper. The total *length* of the

[1] Letter to Vitale Giordano, *Leibnizens mathematische Schriften*, edited by Gerhardt, Section I., Vol. I., page 199.

threads may then serve as measure of the *surfaces*, and the total *area* of the sheets as measure of the *volumes*, and the accuracy of the measurement may be carried to any point we wish. The number of *like equidistant* bodies, if close enough together and of the right form, can just as well furnish the numerical measures of surfaces and solid spaces as the number of identical bodies absolutely covering the surfaces or absolutely filling the spaces. If we cause these bodies to shrink until they become lines (straight lines) or until they become surfaces (planes), we shall obtain the division of surfaces into surface-elements and of spaces into space-elements, and coincidently the customary measurement of surfaces by surfaces and of spaces by spaces.

Cavalieri's defective exposition, which was not adapted to the state of the geometry of his time, has evoked from the historians of geometry some very harsh criticisms of his beautiful and prolific procedure.[1] The fact that a Helmholtz, his critical judgment yielding in an unguarded moment to his fancy, could, in his great youthful work,[2] regard a surface as the sum of the lines (ordinates) contained in it, is merely proof of the great depth to which this original, natural conception reaches, and of the facility with which it reasserts itself.

---

[1] Weissenborn, *Principien der höheren Analysis in ihrer Entwickelung*. Halle, 1856. Gerhardt, *Entdeckung der Analysis*. Halle, 1855, p. 18. Cantor, *Geschichte der Mathematik*. Leipsic, 1892, II. Bd.

[2] Helmholtz, *Erhaltung der Kraft*. Berlin, 1847, p. 14.

The following simple illustration of Cavalieri's method may be helpful to readers not thoroughly conversant with geometry. Imagine a right circular cylinder of horizontal base cut out of a stack of paper sheets resting on a table and conceive inscribed in the cylinder a cone of the same base and altitude. While the sheets cut out by the cylinder are all equal, those forming the cone increase in size as the squares of their distances from the vertex. Now from elementary geometry we know that the

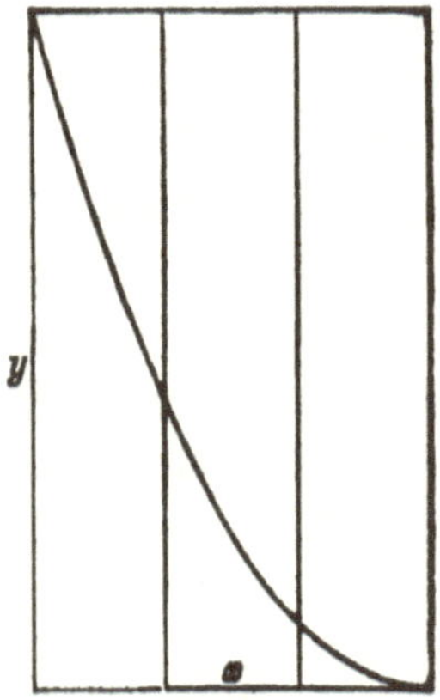

Fig. 3.

volume of such a cone is one-third that of the cylinder. This result may be applied at once to the quadrature of the parabola (Fig. 3). Let a rectangle be described about a portion of a parabola, its sides coinciding with the axis and the tangent to the curve at the origin. Conceiving the rectangle to be covered with a system of threads running parallel to *y*, every thread of the rectangle will be divided into two parts, of which that lying outside the parabola is proportional to x^2. Therefore, the area outside the parabola is to the total area of the rectangle precisely as is the *volume* of the cone to that of the cylinder, viz., as 1 is to 3.

It is significant of the naturalness of Cavalieri's view that the writer of these lines, hearing of the higher geometry when a student at the Gymnasium, but without any training in it, lighted on very similar conceptions,—a performance not attended with any difficulty in the nineteenth century. By the aid of these he made a number of little discoveries, which were of course already long known, found Guldin's theorem, calculated some of Kepler's solids of rotation, etc.

Practical Origin of Geometry.

We have then, first, the general experience that *movable bodies* exist, to which, in spite of their mobility, a certain *spatial constancy* in the sense above described, a permanently *identical property,* must be attributed,—a property which constitutes the foundation of all notions of measurement. But in addition to this there has been gathered instinctively, in the pursuit of the trades and the arts, a considerable variety of *special* experiences, which have contributed their share to the development of geometry. Appearing in part in unexpected form, in part harmonizing with one another, and sometimes, when incautiously applied, even becoming involved in what appears to be paradoxical contradictions, these experiences disturb the course of thought and incite it to the pursuit of the orderly logical connection of these experiences. We shall now devote our attention to some of these processes.

Even though the well known statement of Herodotus[1] were wanting, in which he ascribes the origin of geometry to land-surveying among the Egyptians; and even though the account were totally lost[2] which Eudemus has left regarding the early history of geometry, and which is known to us from an extract in Proclus, it would be impossible for us to doubt that a pre-scientific period of geometry existed. The first geometrical knowledge was acquired accidentally and without design by way of practical experience, and in connection with the most varied employments. It was gained at a time when the scientific spirit, or interest in the interconnection of the experiences in question, was but little developed. This is plain even in our meager history of the beginnings of geometry, but still more so in the history of primitive civilization at large, where technical geometrical appliances are known to have existed at so early and barbaric a day as to exclude absolutely the assumption of scientific effort.

All savage tribes practice the art of weaving, and here, as in their drawing, painting, and wood-cut-

[1] *Herodotus*, II., 109.

[2] James Gow, *A Short History of Greek Mathematics*, Cambridge, 1884, p. 134.

ting, the ornamental themes employed consist of the simplest geometrical forms. For such forms, like the drawings of our children, correspond best to their simplified, typical, schematic conception of the objects which they are desirous of representing and

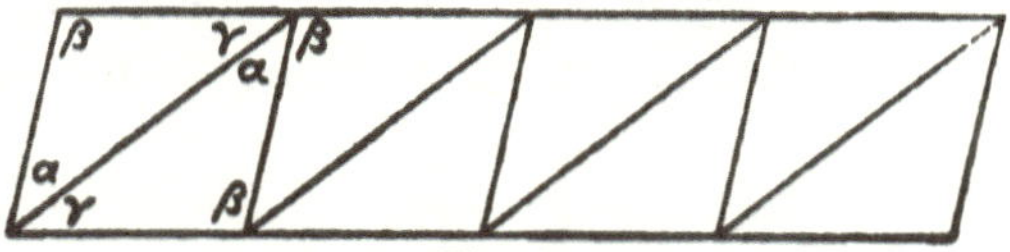

Fig. 4.

it is also these forms that are most easily produced with their primitive implements and lack of manual dexterity. Such an ornament consisting of a series of similarly shaped triangles alternately inverted, or of a series of parallelograms (Fig. 4), clearly suggests the idea, that the sum of the three angles of a triangle, when their vertices are placed together, makes up two right angles. Also this fact could not possibly have escaped the clay and stone workers of Assyria, Egypt, Greece, etc., in constructing their mosaics and pavements from differently colored stones of the same shape. The theorem of the Pythagoreans that the plane space about a point can be completely filled by only three regular polygons, viz., by six equilateral triangles, by four squares, and by three regular hexagons, points to the same source.[1] A like origin of this truth is revealed also in the early Greek method of demonstrat-

[1] This theorem is attributed to the Pythagoreans by Proclus. Cf. Gow, *A Short History of Greek Mathematics*, p. 143, footnote.

ing the theorem regarding the angle-sum of any triangle by dividing it (by drawing the altitude) into two right-angled triangles and completing the rectangles corresponding to the parts so obtained.[1]

The same experiences arise on many other occasions. If a surveyor walk round a polygonal piece of land, he will observe, on arriving at the starting

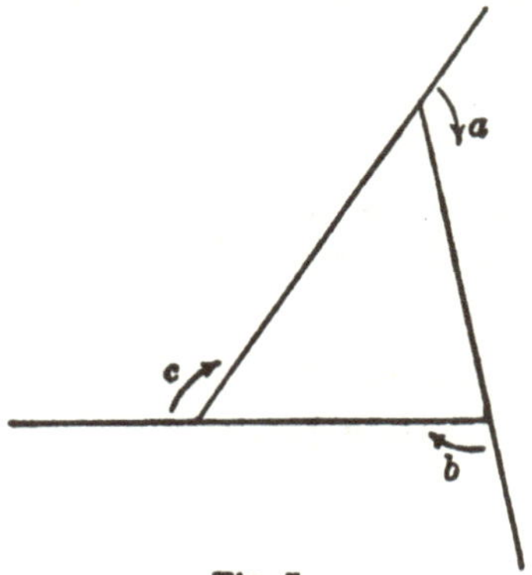

Fig. 5.

point, that he has performed a complete revolution, consisting of four right angles. In the case of a triangle, accordingly, of the six right angles constituting the interior and exterior angles (Fig. 5) there will remain, after subtracting the three exterior angles of revolution, *a, b, c,* two right angles as the sum of the interior angles. This deduction of the theorem was employed by Thibaut,[2] a con-

[1] Hankel, *Geschichte der Mathematik,* Leipsic, 1874, p. 96.

[2] Thibaut, *Grundriss der reinen Mathematik,* Göttingen, 1809, p. 177. The objections which may be raised to this and the following deductions will be considered later. [The same proof is also given by Playfair (1813). See Halsted's translation of Bolyai's *Science Absolute of Space,* p. 67.—*Tr.*]

temporary of Gauss. If a draughtsman draw a triangle by successively turning his ruler round the interior angles, always in the same direction (Fig. 6), he will find on reaching the first side again that if the edge of his ruler lay toward the outside of the triangle on starting, it will now lie toward the inside. In this procedure the ruler has swept out the in-

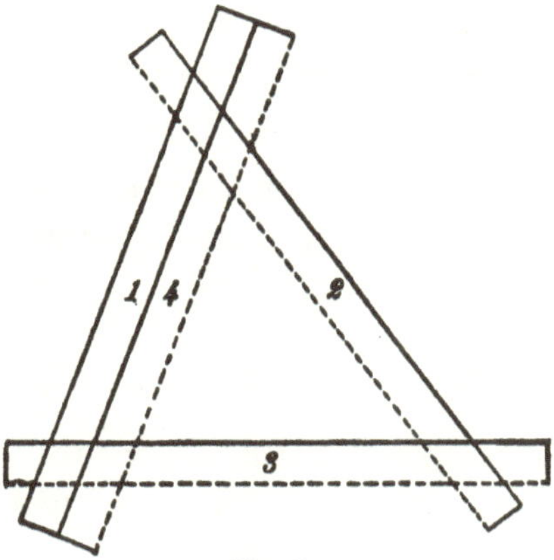

Fig. 6.

terior angles of the triangle in the same direction, and in doing so has performed half a revolution.[1] Tylor[2] remarks that cloth or paper-folding may have led to the same results. If we fold a triangular piece of paper in the manner shown in Fig. 7, we shall obtain a double rectangle, equal in area to one-half the triangle, where it will be seen that the sum of the angles of the triangle coinciding at *a* is two

[1] Noticed by the writer of this article while drawing.

[2] Tylor, *Anthropology, An Introduction to the Study of Man*, etc., German trans., Brunswick, 1883, p. 383.

right angles. Although some very astonishing results may be obtained by paper-folding,[1] it can scarcely be assumed that these processes were *historically* very productive for geometry. The mate-

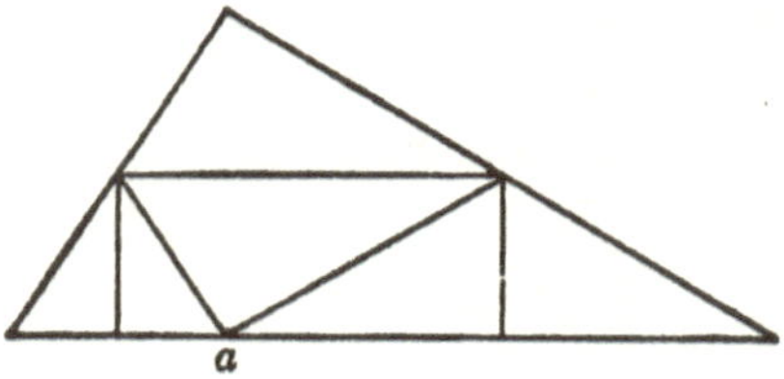

Fig. 7.

rial is of too limited application, and artisans employed with it have too little incentive to exact observation.

Experimental Knowledge of Geometry.

The knowledge that the angle-sum of the plane triangle is equal to a *determinate quantity,* namely, to two right angles, has thus been reached by experience, not otherwise than the law of the lever or Boyle and Mariotte's law of gases. It is true that neither the unaided eye nor measurements with the most delicate instruments can demonstrate *absolutely* that the sum of the angles of a plane triangle is *exactly* equal to two right angles. But the case is precisely the same with the law of the lever and with Boyle's law. All these theorems are therefore idealized and schematized experiences; for real

[1] See, for example, Sundara Row's *Geometric Exercises in Paper-Folding.* Chicago: The Open Court Publishing Co., 1901. —*Tr.*

measurements will always show slight deviations from them. But whereas the law of gases has been proved by further experimentation to be approximate only and to stand in need of modification when the facts are to be represented with great exactness, the law of the lever and the theorem regarding the angle-sum of a triangle have remained in as exact accord with the facts as the inevitable errors of experimenting would lead us to expect; and the same statement may be made of all the consequences that have been based on these two laws as preliminary assumptions.

Equal and similar triangles placed in paving alongside one another with their bases *in one and*

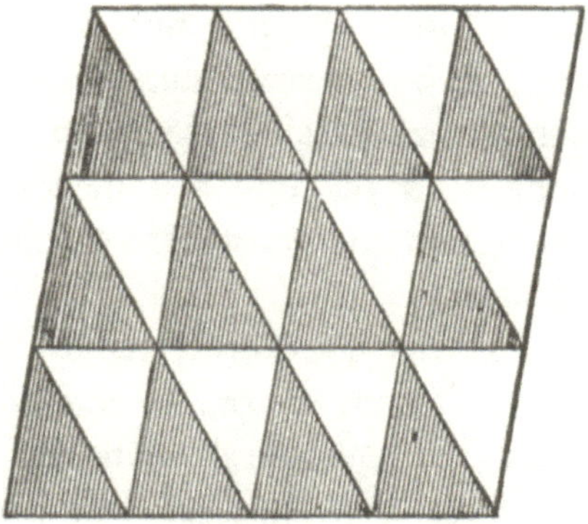

Fig. 8.

the same straight line must also have led to a very important piece of geometrical knowledge. (Fig. 8.) If a triangle be displaced in a plane along a straight line (without rotation), all its points, including those of its bounding lines, will describe equal paths. The same bounding line will furnish, therefore, in any two different positions, a system of two straight lines *equally distant* from one another at all points, and the operation coincidently vouches for the equality of the angles made by the line of displacement on corresponding sides of the two straight lines. The sum of the interior angles on the same side of the line of displacement was consequently determined to be two right angles, and thus Euclid's theorem of parallels was reached. We may add that the possibility of extending a pavement of this kind indefinitely, necessarily lent increased obviousness to this discovery. The sliding of a triangle along a ruler has remained to this day the simplest and most natural method of drawing parallel lines. It is scarcely necessary to remark that the theorem of parallels and the theorem of the angle-sum of a triangle are inseparably connected and represent merely different aspects of the same experience.

The stone masons above referred to must have readily made the discovery that a regular hexagon can be composed of equilateral triangles. Thus resulted immediately the simplest instances of the division of a circle into parts,—namely its division into six parts by the radius, its division into three parts, etc. Every carpenter knows instinctively and almost without reflection that a beam of rectangular symmetric cross-section may, owing to the perfect symmetry of the circle, be cut out from a cylindrical tree-trunk in an infinite number of different ways. The edges of the beam will all lie in the cylindrical surface, and the diagonals of a section will pass through the center. It was in this manner, according to Hankel[1] and Tylor,[2] that the discovery was probably made that all angles inscribed in a semicircle are right angles.

Rôle of Physical Experiences.

A stretched thread furnishes the distinguishing *visualization*[3] of the *straight line.* The straight line is characterized by its physiological simplicity. All its parts induce the *same* sensation of direction; every point evokes the mean of the space-sensations of the neighboring points; every part, however small, is similar to every other part, however great. But, though it has influenced the definitions of many writers,[4] the geometer can accomplish little with this physiological characterization. The visual image must be enriched by physical experience concerning corporeal objects, to be geometrically available. Let a string be fastened by one extremity at *A*, and let its other extremity be passed through a ring fastened at *B*. If we pull on the extremity at *B*, we shall *see* parts of the string which before lay between *A* and *B* pass out at *B*, while at the same

[1] *Loc. cit.*, pp. 206-207.

[2] *Loc. cit.*

[3] *Anschauung.*

[4] Euclid, *Elements*, I., Definition 3.

time the string will approach the form of a straight line. A smaller number of like parts of the string, *identical bodies,* suffices to compose the straight line joining *A* and *B* than to compose a curved line.

It is erroneous to assert that the straight line is recognized as the shortest line *by mere visualization.* It is quite true we can, so far as quality is concerned, reproduce in *imagination* with perfect accuracy and reliability, the simultaneous change of form and length which the string undergoes. But this is nothing more than a reviviscence of a *prior experience with bodies,—an experiment in thought.* The mere *passive contemplation of space* would never lead to such a result. Measurement is experience involving a physical reaction, an experiment of superposition. Visualized or imagined lines having different directions and lengths cannot be applied to one another forthwith. The possibility of such a procedure must be actually experienced with material objects accounted as unalterable. It is erroneous to attribute to animals an instinctive knowledge of the straight line as the shortest distance between two points. If a stimulus excites an animal's attention, and if the animal has so turned that its plane of symmetry passes through the stimulating object, then the straight line is the path of motion *uniquely* determined by the stimulus. This is distinctly shown in Loeb's investigations on the tropisms of animals.

Further, visualization alone does not prove that any two sides of a triangle are together greater than the third side. It is true that if the two sides be laid upon the base by rotation round the vertices of the basal angles, it will be seen by an act of *imagination* alone that the two sides with their free ends moving in arcs of circles will ultimately overlap, thus more than filling up the base. But we should not have attained to this representation had not the procedure been actually witnessed in connection with corporeal objects. Euclid[1] deduces this truth circuitously and artificially from the fact that the greater side of every triangle is opposite to the greater angle. But the source of our knowledge here also is experience,—experience of the motion of the side of a physical triangle; this source has, however, been laboriously concealed by the form of the deduction,—and this not to the enhancement of perspicuity or brevity.

But the properties of the straight line are not exhausted with the preceding empirical truths. If a wire of any arbitrary shape be laid on a board in contact with two upright nails, and slid along so as to be always in contact with the nails, the form and position of the parts of the wire between the nails will be constantly changing. The straighter the wire is, the slighter the alteration will be. A straight wire submitted to the same operation slides *in itself*. Rotated round two of its own fixed points, a crooked wire will keep constantly changing its position, but a straight wire will maintain its position, it will ro-

[1] Euclid, *Elements*, Book I., Prop. 20.

tate within itself.[1] When we define, now, a straight line as the line which is completely determined by two of its points, there is nothing in this *concept* except the *idealization* of the empirical notion derived from the physical experience mentioned,—a notion by no means directly furnished by the physiological act of visualization.

The plane, like the straight line, is physiologically characterized by its simplicity. It appears the same at all parts.[2] Every point evokes the mean of the space-sensations of the neighboring points. Every part, however small, is like every other part, however great. But experiences gained in connection with physical objects are also required, if these properties are to be put to geometrical account. The plane, like the straight line, is physiologically symmetrical with respect to itself, if it coincides with the median plane of the body or stands at right angles to the same. But to discover that symmetry is a *permanent* geometrical property of the plane and the straight line, both concepts must be given as movable, unalterable physical objects. The connection of physiological symmetry with metrical

---

[1] In a letter to Vitale Giordano (*Leibnizens mathematische Schriften, herausgegeben v. Gerhardt, erste Abtheilung, Bd. I., S.* 195-196), Leibnitz makes use of the above-mentioned property of a straight line for its definition. The straight line shares the property of displaceability in itself with the circle and the circular cylindrical spiral. But the property of rotatability within itself and that of being determined by *two* points, are exclusively its own.

[2] Compare Euclid, *Elements* I., Definition 7.

properties also is in need of special metrical demonstration.

Physically a plane is constructed by rubbing three bodies together until three surfaces, *A, B, C,* are obtained, each of which exactly fits the others,—a result which can be accomplished, as Fig. 9 shows, with neither convex nor concave surfaces, but with plane surfaces only. The convexities and concavities are, in fact, removed by the rubbing. Similarly, a truer straight line can be obtained with the aid of an imperfect ruler, by first placing it with its ends against the points *A, B,* then turning it through an angle of 180° out of its plane and again placing it against *A, B,* afterwards taking the *mean* between the two lines so obtained as a more perfect straight line, and repeating the operation with the line last

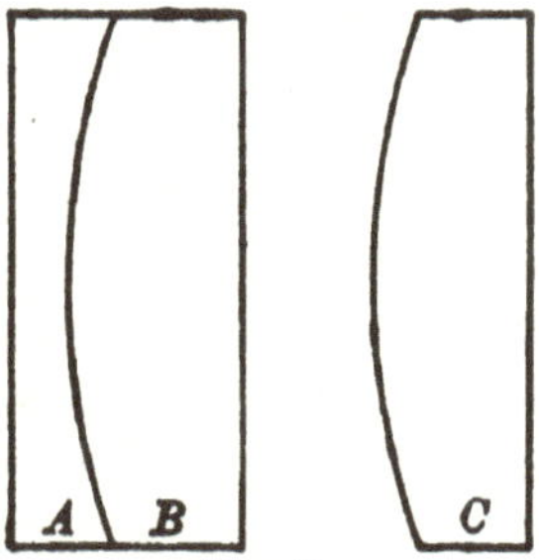

Fig. 9.

obtained. Having by rubbing, produced a plane, that is to say, a surface having the same form at *all points* and on *both* sides, experience furnishes additional results. Placing two such planes one on the other, it will be learned that the plane is *displaceable* into itself, and *rotatable* within itself, just as a straight line is. A thread stretched between any two points in the plane falls entirely within the plane. A piece of cloth drawn tight across any bounded portion of a plane coincides with it. Hence the plane represents the minimum of surface within its boundaries. If the plane be laid on two sharp points, it can still be rotated around the straight line joining the points, but any third point outside of this straight line fixes the plane, that is, determines it completely.

In the letter to Vitale Giordano, above referred to, Leibnitz makes the frankest use of this experience with corporeal objects, when he defines a plane as a surface which divides an unbounded solid into two congruent parts, and a straight line as a line which divides an unbounded plane into two congruent parts.[1]

If attention be directed to the symmetry of the plane with respect to itself, and two points be assumed, one on each side of it, each symmetrical to the other, it will be found that every point in the plane is equidistant from these two points, and Leib-

---

[1] The passage reads literally: "Et difficulter absolvi poterit demonstratio, nisi quis assumat notionem rectæ, qualis est qua ego uti soleo, quod corpore aliquo duobus punctis immotis revoluto locus omnium punctorum quiescentium sit recta, vel saltem quod recta sit linea secans planum interminatum in duas partes congruas; et planum sit superficies secans solidum interminatum in duas partes congruas." For similar definitions, see, for example, Halsted's *Elements of Geometry*, 6th edition. New York, 1895, p. 9.—*T. J. McC.*

nitz's definition of the plane is reached.[1] The uniformity and symmetry of the straight line and the plane are consequences of their being *absolute* minima of length and area respectively. For the boundaries given the minimum must exist, no other collateral condition being involved. The minimum is unique, *single in its kind;* hence the *symmetry* with respect to the bounding points. Owing to the *absoluteness* of the minimum, every portion, however small, again exhibits the same minimal property; hence the uniformity.

EMPIRICAL ORIGIN OF GEOMETRY.

Empirical truths organically connected may make their appearance independently of one another, and doubtless were so discovered long before the fact of their connection was known. But this does not preclude their being afterwards recognized as involved in, and *determined* by, one another, as being *deducible* from one another. For example, supposing we are acquainted with the symmetry and uniformity of the straight line and the plane, we easily deduce that the intersection of two planes is a *straight line,* that any two points of the plane can be joined by a straight line lying wholly within the plane, etc. The fact that only a *minimum* of inconspicuous and unobtrusive experiences is requisite for such deductions should not lure us into the error of regarding this minimum as wholly super-

[1] Leibnitz, *in re* "geometrical characteristic," letter to Huygens, Sept. 8, 1679 (Gerhardt, *loc. cit., erste Abth., Bd. II.,* S. 23).

fluous, and of believing that visualization and reasoning are alone sufficient for the construction of geometry.

Like the concrete visual images of the straight line and the plane, so also our visualizations of the circle, the sphere, the cylinder, etc., are enriched by metrical experiences, and in this manner first rendered amenable to fruitful geometrical treatment. The same economic impulse that prompts our children to retain only the *typical* features in their concepts and drawings, leads us also to the *schematization* and conceptual *idealization* of the images derived from our experience. Although we never come across in nature a perfect straight line or an exact circle, in our thinking we nevertheless designedly abstract from the deviations which thus occur. Geometry, therefore, is concerned with *ideal objects* produced by the schematization of *experiential objects.* I have remarked elsewhere that it is wrong in elementary geometrical instruction to cultivate predominantly the logical side of the subject, and to neglect to throw open to young students the wells of knowledge contained in experience. It is gratifying to note that the Americans who are less dominated than we by tradition, have recently broken with this system and are introducing a sort of experimental geometry as introductory to systematic geometric instruction.*

*See the essays and books of Hanus, Campbell, Speer, Myers, Hall and many others noticed in the reviews of *School Science and Mathematics* (Chicago) during the last few years.—*T. J. McC.*

TECHNICAL AND SCIENTIFIC DEVELOPMENT OF GEOMETRY.

No sharp line can be drawn between the instinctive, the technical, and the scientific acquisition of geometric notions. Generally speaking, we may say, perhaps, that with division of labor in the industrial and economic fields, with increasing employment with very definite objects, the instinctive acquisition of knowledge falls into the background, and the technical begins. Finally, *when measurement becomes an aim and business in itself*, the *connection* obtaining between the various operations of measuring acquires a powerful *economic* interest, and we reach the period of the scientific development of geometry, to which we now proceed.

The insight that the measures of geometry depend on one another, was reached in divers ways. After surfaces came to be measured by surfaces, further progress was almost inevitable. In a parallelogrammatic field permitting a division into equal partial parallelogrammatic fields so that n rows of partial fields each containing m fields lay alongside one another, the counting of these fields was unnecessary. By multiplying together the numbers measuring the sides, the area of the field was found to be equal to mn such fields, and the area of each of the two triangles formed by drawing the diagonal was readily discovered to be equal to $\frac{mn}{2}$ such fields. This was the first and simplest application of arithmetic to geometry. Coincidently, the dependence of measures of area on other measures, linear and angular, was discovered. The area of a rectangle was found to be larger than that of an oblique parallelogram having sides of the same length; the area, consequently, depended not only on the length of the sides, but also on the angles. On the other hand, a rectangle constructed of strips of wood running parallel to the base, can, as is easily seen, be converted by displacement into any parallelogram of the same height and base without altering its area. Quadrilaterals having their sides given are still undetermined in their angles, as every carpenter knows. He adds diagonals, and converts his quadrilateral into triangles, which, the sides being given, are rigid, that is to say, are unalterable as to their angles also.

With the perception that measures were dependent on one another, the real problem of geometry was introduced. Steiner has aptly and justly entitled his principal work "Systematic Development of the Dependence of Geometrical Figures on One Another."[1] In Snell's original but unappreciated treatise on Elementary Geometry, the problem in question is made obvious even to the beginner.[2]

A plane physical triangle is constructed of wires. If one of the sides be rotated around a vertex, so as to increase the interior angle at that point, the side

[1] J. Steiner, *Systematische Entwicklung der Abhängigkeit der geometrischen Gestalten von einander.*

[2] Snell, *Lehrbuch der Geometrie*, Leipsic, 1869.

moved will be seen to change its position and the side opposite to grow *larger* with the angle. *New* pieces of wire besides those before present will be required to complete the last-mentioned side. This and other similar experiments can be repeated in thought, but the mental experiment is never anything more than a copy of the physical experiment. The mental experiment would be impossible if physical experience had not antecedently led us to a knowledge of *spatially unalterable physical bodies*,[1] —to the concept of measure.

The Geometry of the Triangle.

By experiences of this character, we are conducted to the truth that of the six metrical magnitudes discoverable in a triangle (three sides and three angles) three, including at least *one* side, suffice to determine the triangle. If *one* angle only be given among the parts determining the triangle, the angle in question must be either the angle included by the given sides, or that which is opposite to the greater side,—at least if the determination is to be *unique*. Having reached the perception that a triangle is determined by three sides and that its form is independent of its position, it follows that in an equilateral triangle all three angles and in an isosceles triangle the two angles opposite the equal sides, must be equal, in

[1] The whole construction of the Euclidean geometry shows traces of this foundation. It is still more conspicuous in the "geometric characteristic" of Leibnitz already mentioned. We shall revert to this topic later.

whatever manner the angles and sides may depend on one another. This is logically certain. But the empirical foundation on which it rests is for that reason not a whit more superfluous than it is in the analogous cases of physics.

The mode in which the sides and angles depend on one another is, naturally, first recognized in special instances. In computing the areas of rectangles and of the triangles formed by their diagonals, the fact must have been noticed that a rectangle having sides 3 and 4 units in length gives a right-angled triangle having sides, 3, 4, and 5 units in length. Rectangularity was thus shown to be connected with a definite, rational ratio between the sides. The knowledge of this truth was employed to stake off right angles, by means of three connected ropes respectively 3, 4, and 5 units in length.[1] The equation $3^2 + 4^2 = 5^2$, the analogue of which was proved to be valid for all right-angled triangles having sides of lengths a, b, c (the general formula being $a^2 + b^2 = c^2$), now riveted the attention. It is well known how profoundly this relation enters into metrical geometry, and how all indirect measurements of distance may be traced back to it. We shall endeavor to disclose the foundation of this relation.

It is to be remarked first that neither the Greek geometrical nor the Hindu arithmetical deductions of the so-called *Pythagorean Theorem* could avoid the consideration of areas. One essential point on

[1] Cantor, *Geschichte der Mathematik*, Leipsic, 1880. I., pp. 53, 56.

which all the deductions rest and which appears more or less distinctly in different forms in all of them, is the following. If a triangle, *a, b, c* (Fig. 10) be slid along a short distance in its own plane, it is as-

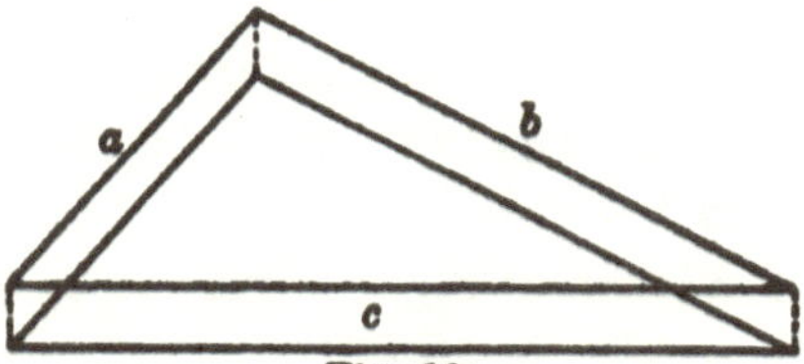

Fig. 10.

sumed that the space which it leaves behind is compensated for by the new space on which it enters. That is to say, the area swept out by *two* of the sides during the displacement is equal to the area swept out by the *third* side. The basis of this conception is the assumption of the *conservation of the area* of the triangle. If we consider a surface as a body of very minute but unvarying thickness of third dimension (which for that reason is uninfluential in the present connection), we shall again have the *conservation of the volume of bodies* as our fundamental assumption. The same conception may be applied to the translation of a tetrahedron, but it does not lead in this instance to new points of view. Conservation of volume is a property which rigid and liquid bodies possess in common, and was idealized by the old physics as *impenetrability*. In the case of rigid bodies, we have the additional attribute that the distances between all the parts are preserved, while in the case of liquids, the properties of rigid bodies exist only for the smallest time and space elements.

If an oblique-angled triangle having the sides *a*, *b*, and *c* be displaced in the direction of the side *b*, only *a* and *c* will, by the principle above stated, describe equivalent parallelograms, which are alike in an equal pair of parallel sides on the same parallels. If *a* make with *b* a right angle, and the triangle be displaced at right angles to *c*, the distance *c*, the side *c* will describe the square c^2, while the two other sides will describe parallelograms the combined areas of which are equal to the area of the

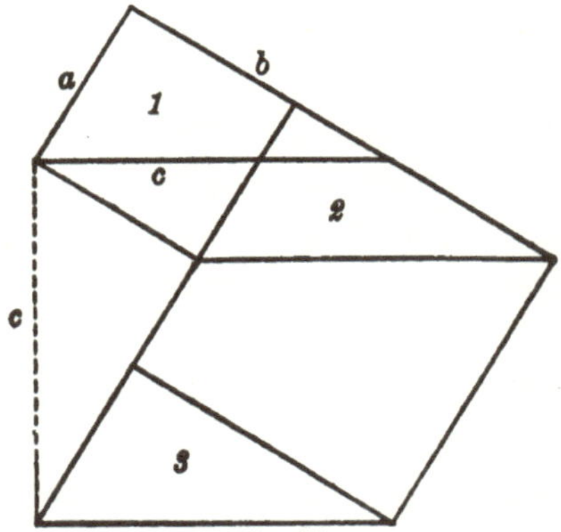

Fig. 11.

square. But the two parallelograms are, by the observation which just precedes, equivalent respectively to a^2 and b^2,—and with this the Pythagorean theorem is reached.

The same result may also be attained (Fig. 11) by first sliding the triangle a distance *a* at right angles to *a*, and then a distance *b* at right angles to

b, where $a^2 + b^2$ will be equal to the sum of the surfaces swept out by c, which is obviously c^2. Taking an oblique-angled triangle, the same procedure just as easily and obviously gives the more general proposition, $c^2 = a^2 + b^2 - 2ab\cos\gamma$.

The dependence of the third side of the triangle on the two other sides is accordingly determined by the area of the enclosed triangle; or, in our conception, by a condition involving *volume*. It will also be directly seen that the equations in question express relations of area. It is true that the angle included between two of the sides may also be regarded as determinative of the third side, in which case the equations will aparently assume an entirely different form.

Let us look a little more closely at these different measures. If the extremities of two straight lines of lengths a and b meet in a point, the length of the line c joining their free extremities will be included between definite limits. We shall have $c \overset{=}{<} a + b$, and $c \overset{=}{>} a - b$. Visualization alone cannot inform us of this fact; we can learn it only from *experimenting in thought*,—a procedure which both reposes on physical experience and reproduces it. This will be seen by holding a fast, for example, and turning b, first, until it forms the prolongation of a, and, secondly, until it coincides with a. A straight line is primarily a unique concrete image characterized by physiological properties,—an image which we have obtained from a *physical* body of a definite specific character, which in the form of a string or wire of indefinitely small but constant thickness interposes a *minimum of volume* between the positions of its extremities,—which can be accomplished only in *one uniquely-determined* manner. If several straight lines pass through a point, we distinguish between them *physiologically* by their directions. But in *abstract space* obtained by metrical experiences with physical objects, differences of direction do not exist. A straight line passing through a point can be completely determined in abstract space only by assigning a second *physical* point on it. To define a straight line as a line which is constant in direction, or an angle as a *difference between directions,* or parallel straight lines as straight lines having the *same* direction, is to define these concepts *physiologically.*

The Measurement of the Angle.

Different methods are at our disposal when we come to characterize or determine *geometrically* angles which are *visually* given. An angle is determined when the distance is assigned between any two fixed points lying each on a separate side of the angle outside the point of intersection. To render the definition uniform, points situated at the same fixed and invariable distance from the vertex might be chosen. The inconvenience that then equimultiples of a given angle placed alongside one another in the same plane with their vertices coincident, would not be measured by the same equimultiples of the distance between those points, is the reason that this method of determining angles was not introduced into elementary geometry.[1] A simpler measure, a simpler characterization of an angle, is obtained by taking the aliquot part of the *circumference* or the *area* of a circle which the angle intercepts when laid in the plane of the circle with its vertex at the center. The convention here involved is more convenient.[2]

In employing an arc of a circle to determine an angle, we are again merely measuring a volume,—viz., the volume occupied by a body of simple definite form introduced between two points on the arms of the angle equidistant from the vertex. But a circle can be characterized by simple rectilinear distances. It is a matter of perspicuity, of immediacy, and of the facility and convenience resulting therefrom, that two measures, viz., the rectilinear measure of length and the angular measure, are principally employed as fundamental measures, and that the others are derived from them. It is in no sense necessary. For example (Fig. 12), it is possible without a special angular measure to determine the straight line that cuts another straight line at right angles by making all its points equidistant from two points in the first straight line lying at equal distances from the point of intersection. The

[1] A closely allied principle of measurement is, however, applied in trigonometry.

[2] So also the superficial portion of a sphere intercepted by the including planes is used as the measure of a solid angle.

bisector of an angle can be determined in a quite similar manner, and by continued bisection an angular unit can be derived of any smallness we wish. A straight line *parallel* to another straight

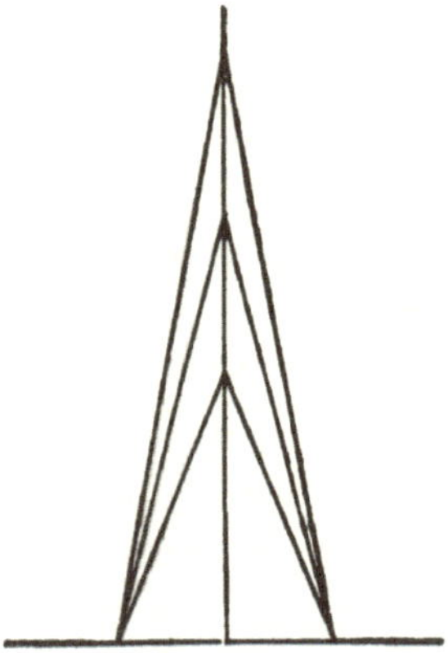

Fig. 12.

line can be defined as one, all of whose points can be translated by congruent curved or *straight* paths into points of the first straight line.[1]

Length as the Fundamental Measure.

It is quite possible to start with the straight length *alone* as our fundamental measure. Let a fixed physical point a be given. Another point, m, has the distance r_a from the first point. Then this last point can still lie in any part of the spherical surface described about a with radius r_a. If we know still a second fixed point b, from which m is removed by the distance r_b, the triangle abm will

[1] If this form had been adopted, the doubts as to the Euclidean theorem of parallels would probably have risen much later.

be rigid, determined; but *m* can still revolve round in the circle described by the rotation of the triangle around the axis *ab*. If now the point *m* be held fast in any position, then also the whole rigid body to which the three points in question, *a, b, m,* belong will be fixed.

A point *m* is spatially determined, accordingly, by the distances r_a, r_b, r_c from at least three fixed points in space, *a, b, c*. But this determination is still not unique, for the pyramid with the edges r_a, r_b, r_c, in the vertex of which *m* lies, can be constructed as well on the one as on the other side of the plane *a, b, c*. If we were to fix the side, say by a special sign, we should be resorting to a *physiological* determination, for *geometrically* the two sides of the plane are not different. If the point *m* is to be uniquely determined, its distance, r_d, from a fourth point, *d*, lying *outside* the plane *abc*, must in addition be given. Another point, *m'*, is determined with like completeness by four distances, r'_a, r'_b, r'_c, r'_d. Hence, the distance of *m* from *m'* is also given by this determination. And the same holds true of any number of other points as severally determined by four distances. Between four points $\frac{4(4-1)}{1\cdot 2} = 6$ distances are conceivable, and precisely this number must be given to determine the form of the point complex. For $4 + z = n$ points, $6 + 4z$ or $4n - 10$ distances are needed for the determination, while a still larger number, viz.,

$\frac{n(n-1)}{1 \cdot 2}$ distances exist, so that the excess of the distances is also coincidently determined.*

If we start from *three* points and prescribe that the distances of all points to be further determined shall hold for one side only of the plane determined by the three points, then $3n-6$ distances will suffice to determine the form, magnitude, and position of a system of n points with respect to the three initial points. But if there be no condition as to the side of the plane to be taken,—a condition which involves sensuous and physiological, but not abstract metrical characteristics,—the system of points, instead of the intended form and position, may assume that symmetrical to the first, or be combined of the points of both. *Symmetric* geometrical figures are, owing to our symmetric *physiological organization*, very easily taken to be identical, whereas *metrically* and *physically* they are entirely different. A screw with its spiral winding to the right and one with its spiral winding to the left, two bodies rotating in contrary directions, etc., appear very much alike to the eye. But we are for this reason not permitted to regard them as geometrically or physically equivalent. Attention to this fact would avert many paradoxical questions. Think only of the trouble that such problems gave Kant!

*For an interesting attempt to found both the Euclidean and non-Euclidean geometrics on the pure notion of distance, see De Tilly, "Essai sur les principes fondamentaux de la géométrie et de la mécanique" (*Mémoires de la Société de Bordeaux*, 1880).

Sensuous physiological attributes are determined by relationship to *our body,* to a corporeal system of *specific* constitution; while metrical attributes are determined by relations to the world of physical bodies *at large.* The latter can be ascertained only by experiments of coincidence,—by measurements.

Volume the Basis of Measurement.

As we see, every geometrical measurement is at bottom reducible to measurements of *volumes,* to the *enumeration of bodies.* Measurements of lengths, like measurements of areas, repose on the comparison of the volumes of very thin strings, sticks, and leaves of constant thickness. This is not at variance with the fact that measures of area may be *arithmetically* derived from measures of length, or solid measures from measures of length alone, or from these in combination with measures of area. This is merely proof that *different* measures of volume are dependent on one another. To ascertain the forms of this interdependence is the *fundamental object of geometry,* as it is the province of arithmetic to ascertain the manner in which the various numerical operations, or ordinative activities of the mind, are connected together.

The Visual Sense in Geometry.

It is extremely probable that the experiences of the visual sense were the cause of the *rapidity* with which geometry developed. But our great familiarity with the properties of rays of light gained from the present advanced state of optical technique, should not mislead us into regarding our *experimental knowledge of rays of light* as the principal foundation of geometry. Rays of light in dust or smoke-laden air furnish admirable *visualizations* of straight lines. But we can derive the *metrical properties* of straight lines from rays of light just as little as we can derive them from *imaged* straight lines. For this purpose experiences with *physical* objects are absolutely necessary. The *rope-stretching* of the practical geometers is certainly older than the use of the theodolite. But once knowing the physical straight line, the ray of light furnishes a very distinct and handy means of reaching new points of view. A blind man could scarcely have invented modern synthetic geometry. But the oldest and the most powerful of the experiences lying at the basis of geometry are just as accessible to the blind man, through his sense of touch, as they are to the person who can see. Both are acquainted with the spatial *permanency of bodies* despite their *mobility;* both acquire a conception of *volume* by *taking hold* of objects. The creator of primitive geometry disregards, first instinctively and then intentionally and consciously, those physical properties that are unessential to his operations and that for the moment do not concern him. In this manner, and by gradual growth, the idealized concepts of geometry arise on the basis of experience.

Various Sources of Our Geometric Knowledge.

Our geometrical knowledge is thus derived from various sources. We are *physiologically* acquainted, from direct visual and tactual contact, with many and various spatial forms. With these are associated physical (*metrical*) experiences (involving comparison of the space-sensations evoked by different bodies under the same circumstances), which experiences are in their turn also but the expressions of other relations obtaining between sensations. These diverse orders of experience are so intimately interwoven with one another that they can be separated only by the most thoroughgoing scrutiny and analysis. Hence originate the widely divergent views concerning geometry. Here it is based on pure visualization (*Anschauung*), there on physical experience, according as the one or the other factor is overrated or disregarded. But both factors entered into the development of geometry and are still active in it to-day; for, as we have seen, geometry by no means exclusively employs purely metrical concepts.

If we were to ask an unbiased, candid person under what form he pictured space, referred, for example, to the Cartesian system of co-ordinates, he would doubtless say: I have the image of a system of rigid (form-fixed), transparent, penetrable, contiguous cubes, having their bounding surfaces marked only by nebulous visual and tactual percepts,—a species of phantom cubes. Over and through these phantom constructions the real bodies or their phantom counterparts move, conserving their spatial permanency (as above defined), whether we are concerned with practical or theoretical geometry, or phoronomy. Gauss's famous investigation of curved surfaces, for instance, is really concerned with the application of infinitely thin laminate and hence flexible bodies to one another. That diverse orders of experience have co-operated in the formation of the fundamental conceptions under consideration, cannot be gainsaid.

The Fundamental Facts and Concepts.

Yet, varied as the special experiences are from which geometry has sprung, they may be reduced to a minimum of facts: Movable bodies exist having definite spatial permanency,—viz., rigid bodies exist. But the movability is characterized as follows: we draw from a point three lines not all in the same plane but otherwise undetermined. By three movements along these straight lines any point can be reached from any other. Hence, three measurements or dimensions, physiologically and metrically characterized as the simplest, are sufficient for all spatial determinations. These are the fundamental facts.[1]

The physical metrical experiences, like all experi-

[1] The historical development of this conception will be considered in another place.

ences forming the basis of experimental sciences, are conceptualized,—idealized. The *need* of representing the facts by simple perspicuous concepts under easy logical control, is the reason for this. Absolutely rigid, spatially invariable bodies, perfect straight lines and planes, no more exist than a perfect gas or a perfect liquid. Nevertheless, deferring the consideration of the deviations, we prefer to work, and we also work more readily, with these concepts than with others that conform more closely to the actual properties of the objects. *Theoretical* geometry does not even need to consider these deviations, inasmuch as it assumes objects that fulfil the requirements of the theory absolutely, just as theoretical physics does. But in *practical* geometry, where we are concerned with actual objects, we are obliged, as in practical physics, to consider the deviations from the theoretical assumptions. But geometry has still the advantage that every deviation of its objects from the assumptions of the theory *which may be detected* can be *removed;* whereas physics for obvious reasons cannot construct more perfect gases than actually exist in nature. For, in the latter case, we are concerned not with a *single* arbitrarily constructible spatial property alone, but with a relation (occurring in nature and independent of our will) between pressure, volume, and temperature.

The choice of the concepts is suggested by the facts; yet, seeing that this choice is the outcome of our *voluntary* reproduction of the facts in thought, some free scope is left in the matter. The importance of the concepts is estimated by their range of application. This is why the concepts of the straight line and the plane are placed in the foreground, for every geometrical object can be split up with sufficient approximateness into elements bounded by planes and straight lines. The particular properties of the straight line, plane, etc., which we decide to emphasize, are matters of our own free choice, and this truth has found expression in the various definitions that have been given of the same concept.[1]

Experimenting in Thought.

The fundamental truths of geometry have thus, unquestionably, been derived from physical experience, if only for the reason that our visualizations and sensations of space are absolutely inaccessible to measurement and cannot possibly be made the subject of metrical experience. But it is no less indubitable that when the relations connecting our visualizations of space with the simplest metrical experiences have been made familiar, then geometrical facts can be reproduced with great facility and certainty in the imagination alone,—that is by purely *mental experiment*. The very fact that a continuous change in our space-sensation corresponds to a continuous metrical change in physical bodies, enables

---

[1] Compare, for example, the definitions of the straight line given by Euclid and by Archimedes.

us to ascertain by imagination alone the particular metrical elements that depend on one another. Now, if such metrical elements are observed to enter different constructions having different positions in precisely the same manner, then the metrical results will be regarded as *equal*. The case of the isosceles and equilateral triangles, above mentioned, may serve as an example. The *geometric* mental experiment has advantage over the physical, only in the respect that it can be performed with far simpler experiences and with such as have been more easily and almost unconsciously acquired.

Our sensuous imagings and visualizations of space are *qualitative,* not quantitative nor metrical. We derive from them coincidences and differences of extension, but never real magnitudes. Conceive, for example, Fig. 13, a coin rolling clockwise down and around the rim of another fixed coin of the same size, without sliding. Be our imagination as vivid as it will, it is impossible by a pure feat of reproductive imagery alone, to determine here the angle described in a full revolution. But if it be considered that at the beginning of the motion the radii *a, a'* lie in one straight line, but that after a *quarter* revolution the radii *b, b'* lie in a straight line, it will be seen at once that the radius *a'* now points vertically upwards and has consequently performed *half* a revolution. The *measure* of the revolution is obtained from metrical concepts, which fix idealized experiences on definite physical objects, but the *direction* of the revolution is retained in the *sensuous* imagination. The metrical concepts simply determine that in equal circles equal angles are subtended by equal arcs, that the radii to the point of contact lie in a straight line, etc.

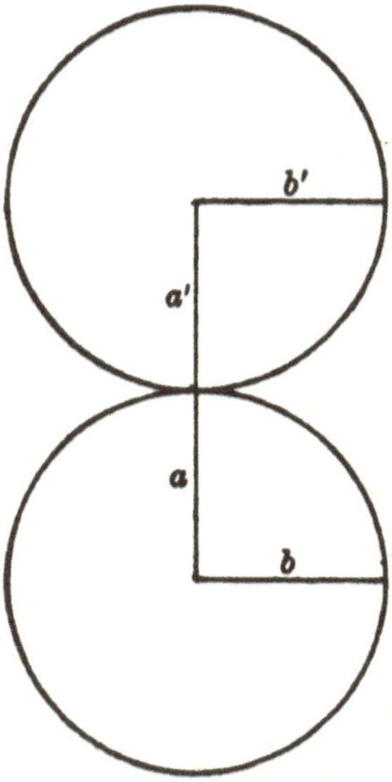

Fig. 13.

If I picture to myself a triangle with one of its angles increasing, I shall also see the side opposite the angle increasing. The impression thus arises that the interdependence in question follows *a priori* from a feat of imagination alone. But the imagination has here merely reproduced a fact of experience. Measure of angle and measure of side are *two physical* concepts applicable to *the same* fact,—concepts that have grown so familiar to us that they have come to be regarded as merely *two* different attributes of the *same* imaged group of facts, and hence appear as linked together of sheer necessity. Yet we should never have acquired these concepts without physical experience.

The combined action of the sensuous imagination with idealized concepts derived from experience is apparent in every geometrical deduction. Let us consider, for example, the simple theorem that the perpendicular bisectors of the sides of a triangle *ABC* meet in a common point. Experiment and imagination both doubtless led to the theorem. But the more carefully the construction is executed, the more one becomes convinced that the third perpendicular does not pass exactly through the point of intersection of the first two, and that in any actual construction, therefore, three points of intersection will be found closely adjacent to one another. For in reality neither perfect straight lines nor perfect perpendiculars can be drawn; nor can the latter be erected exactly at the midpoints; and so on. *Only* on the assumption of these *ideal* conditions does the perpendicular bisector of *AB* contain all points equally distant from *A* and *B*, and the perpendicular bisector of *BC* all points equidistant from *B* and *C*. From which it follows that the point of intersection of the two is equidistant from *A*, *B*, and *C*, and by reason of its equidistance from *A* and *C* is also a point of the *third* perpendicular bisector, of *AC*. The theorem asserts therefore that the more accurately the assumptions are fulfilled the more nearly will the three points of intersection coincide.

Kant's Theory.

The importance of the combined action of the sensuous imagination [viz., of the *Anschauung* or intuition so called] and of concepts, will doubtless have been rendered clear by these examples. Kant says: "Thoughts without contents are empty, intuitions without concepts are blind."[1] Possibly we might more appropriately say: "Concepts without intuitions are blind, intuitions without concepts are lame." For it would appear to be not so absolutely correct to call intuitions [viz., sensuous images] blind and concepts empty. When Kant further says that "there is in every branch of natural knowledge only so much science as there is mathematics contained in it,"[2] one might possibly also assert of all sciences, *including* mathematics, "that they are only in so far sciences as they operate with concepts." For our logical mastery extends only to those concepts of which we have ourselves determined the contents.

The Present Form of Geometry.

The two facts that bodies are rigid and movable would be sufficient for an understanding of any geometrical fact, no matter how complicated,—sufficient, that is to say, to derive it from the two facts

---

[1] *Kritik der reinen Vernunft*, 1787, p. 75. Max Müller's translation, 2nd ed., 1896, p. 41.

[2] *Metaphysische Anfangsgründe der Naturwissenschaft. Vorwort.*

mentioned. But geometry is obliged, both in its own interests and in its rôle as an auxiliary science, as well as in the pursuit of practical ends, to answer questions that *recur repeatedly in the same form.* Now it would be uneconomical, in such a contingency, to begin each time with the most elementary facts and to go to the bottom of each new case that presented itself. It is preferable, rather, to select a few simple, familiar, and indubitable theorems, in our choice of which caprice is by no means excluded,[1] and to formulate from these, *once for all,* for application to practical ends, general propositions answering the questions that most frequently recur. From this point of view we understand at once the *form* geometry has assumed,—the emphasis, for example, that it lays upon its propositions concerning triangles. For the purpose designated it is desirable to collect the most general possible propositions having the widest range of application. From history we know that propositions of this character have been obtained by embracing various special cases of knowledge under single general cases. We are forced even today to resort to this procedure when we treat the relationship of two geometrical figures, or when the different special cases of form and position compel us to modify our modes of deduction. We may cite as the most familiar instance of this in elementary geometry, the

[1] Zindler. *Zur Theorie der mathematischen Erkenntniss. Sitzungsberichte der Wiener Akademie. Philos-histor. Abth.* Bd. 118. 1889.

mode of deducing the relation obtaining between angles at the centre and angles at the circumference.

UNIVERSAL VALIDITY.

Kroman[1] has put the question, Why do we regard a demonstration made with a special figure (a special triangle) as universally valid for all figures? and finds his answer in the supposition that we are able by rapid variations to impart all possible forms to the figure in thought and so convince ourselves of the admissibility of the same mode of inference in all special cases. History and introspection declare this idea to be in all essentials correct. But we may not assume, as Kroman does, that in each special case every individual student of geometry acquires this perfect comprehension "with the rapidity of lightning," and reaches immediately the lucidity and intensity of geometric conviction in question. Frequently the required operation is absolutely impracticable, and errors prove that in other cases it was actually not performed but that the inquirer rested content with a conjecture based on analogy.[2]

But that which the individual does not or cannot achieve in a jiffy, he may achieve in the course of his life. Whole generations labor on the verification of geometry. And the conviction of its certitude is unquestionably strengthened by their collec-

---

[1] *Unsere Naturerkenntniss.* Copenhagen, 1883, pp. 74 et seq.

[2] Hoelder, *Anschauung und Denken in der Geometrie*, p. 12.

tive exertions. I once knew an otherwise excellent teacher who compelled his students to perform all their demonstrations with *incorrect* figures, on the theory that it was the *logical* connection of the concepts, not the figure, that was essential. But the experiences imbedded in the concepts cleave to our sensuous images. Only the actually visualized or imaged figure can tell us what particular concepts are to be employed in a given case. The method of this teacher is admirably adapted for rendering palpable the degree to which logical operations participate in reaching a given perception. But to employ it habitually is to miss utterly the truth that abstract concepts draw their ulitmate power from sensuous sources.

SPACE AND GEOMETRY FROM THE POINT OF VIEW OF PHYSICAL INQUIRY.[1]

Our notions of space are rooted in our *physiological* organism. Geometric concepts are the product of the idealization of *physical* experiences of space. Systems of geometry, finally, originate in the *logical* classification of the conceptual materials so obtained. All three factors have left their indubitable traces in modern geometry. Epistemological inquiries regarding space and geometry accordingly concern the physiologist, the psychologist, the physicist, the mathematician, the philosopher, and the logician alike, and they can be gradually carried to their definitive solution only by the consideration of the widely disparate points of view which are here offered.

Awakening in early youth to full consciousness, we find ourselves in possession of the notion of a *space* surrounding and encompassing our body, in which space move divers *bodies*, now altering and

---

[1]I shall endeavor in this essay to define my attitude as a physicist toward the subject of metageometry so called. Detailed geometric developments will have to be sought in the sources. I trust, however, that by the employment of illustrations which are familiar to every one I have made my expositions as popular as the subject permitted.

now retaining their size and shape. It is impossible for us to ascertain how this notion has been begotten. Only the most thoroughgoing analysis of experiments purposefully and methodically performed has enabled us to conjecture that inborn idiosyncracies of the body have coöperated to this end with simple and crude experiences of a purely physical character.

Sensational and Locative Qualties.

An object seen or touched is distinguished not only by a *sensational quality* (as "red," "rough," "cold," etc.), but also by a *locative quality* (as "to the left," "above," "before," etc.). The sensational quality may remain the same, while the locative quality continuously changes; that is, the same sensuous object may move in space. Phenomena of this kind being again and again induced by physico-physiological circumstances, it is found that however varied the accidental sensational qualities may be, the same order of locative qualities invariably occurs, so that the latter appear perforce as a fixed and permanent system or register in which the sensational qualities are entered and classified. Now, although these qualities of sensation and locality can be excited only in conjunction with one another, and can make their appearance only concomitantly, the impression nevertheless easily arises that the more familiar system of locative qualities is given antecedently to the sensational qualities (Kant).

Extended objects of vision and of touch consist

of more or less distinguishable sensational qualities, conjoined with adjacent distinguishable, continuously graduated locative qualities. If such objects move, particularly in the domain of our hands, we perceive them to shrink or swell (in whole or in part), or we perceive them to remain the same; in other words, the contrasts characterizing their bounding locative qualities change or remain constant. In the latter case, we call the objects rigid. By the recognition of permanency as coincident with spatial displacement, the various constituents of our intuition of space are rendered *comparable* with one another,—at first in the *physiological* sense. By the comparison of different bodies with one another, by the introduction of *physical* measures, this comparability is rendered quantitative and more exact, and so transcends the limitations of individuality. Thus, in the place of an individual and non-transmittable intuition of space are substituted the universal concepts of geometry, which hold good for all men. Each person has his own individual intuitive space; geometric space is common to all. Between the space of intuition and *metric* space, which contains physical experiences, we must distinguish sharply.

Riemann's Physical Conception of Geometry.

The need of a thoroughgoing epistemological elucidation of the foundations of geometry induced Riemann,[1] about the middle of the century just

[1] *Ueber die Hypothesen, welche der Geometrie zu Grunde liegen.* Göttingen, 1867.

closed, to propound the question of the nature of space; the attention of Gauss, Lobachévski, and Bolyai having before been drawn to the empirically hypothetical character of certain of the fundamental assumptions of geometry. In characterizing space as a special case of a multiply-extended "magnitude," Riemann had doubtless in mind some geometric construct, which may in the same manner be imagined to fill all space,—for example, the system of Cartesian co-ordinates. Riemann further asserts that "the propositions of geometry cannot be deduced from general conceptions of magnitude, but that the peculiar properties by which space is distinguished from other conceivable triply-extended magnitudes can be derived from experience only....These facts, like all facts, are in no wise necessary, but possess empirical certitude only,—they are hypotheses." Like the fundamental assumptions of every natural science, so also, on Riemann's theory, the fundamental assumptions of geometry, to which experience has led us, are merely *idealizations* of experience.

In this physical conception of geometry, Riemann takes his stand on the same ground as his master Gauss, who once expressed the conviction that it was impossible to establish the foundations of geometry entirely *a priori*,[1] and who further asserted that "we must in humility confess that if number is exclusively a product of the mind, space

[1] *Brief von Gauss an Bessel*, 27. Januar 1829.

possesses in addition a reality outside of our mind, of which reality we cannot fully dictate *a priori* the laws."[1]

Analogies of Space with Colors.

Every inquirer knows that the knowledge of an object he is investigating is materially augmented by *comparing* it with related objects. Quite naturally therefore Riemann looks about him for objects which offer some analogy to space. Geometric space is defined by him as a triply-extended continuous manifold, the elements of which are the points determined by every possible three co-ordinate values. He finds that "the places of sensuous objects and colors are probably the only concepts [*sic*] whose modes of determination form a multiply-extended manifold." To this analogy others were added by Riemann's successors and elaborated by them, but not always, I think, felicitously.[2]

[1] *Brief von Gauss an Bessel.* April 9, 1830.—The phrase, "Number is a product or creation of the mind," has since been repeatedly used by mathematicians. Unbiased psychological observation informs us, however, that the formation of the concept of number is just as much initiated by experience as the formation of geometric concepts. We must at least know that virtually *equivalent* objects exist in multiple and unalterable form before concepts of number can originate. Experiments in counting also play an important part in the development of arithmetic.

[2] When acoustic pitch, intensity, and *timbre*, when chromatic tone, saturation, and luminous intensity are proposed as analogues of the three dimensions of space, few persons will be satisfied. *Timbre*, like chromatic tone, is dependent on several variables. Hence, if the analogy has any meaning whatever, several dimensions will be found to correspond to *timbre* and chromatic tone.

Comparing *sensation* of space with *sensation* of color, we discover that to the continuous series "above and below," "right and left," "near and far," correspond the three sensational series of mixed colors, black-white, red-green, blue-yellow. The system of sensed (seen) places is a triple continuous manifold like the system of color-sensations. The objection which is raised against this analogy, viz., that in the first instance the three variations (dimensions) are homogeneous and interchangeable with one another, while in the second instance they are heterogeneous and not interchangeable, does not hold when space-*sensation* is compared with color-*sensation*. For from the psycho-physiological point of view "right and left" as little permit of being interchanged with "above and below" as do red and green with black and white. It is only when we compare *geometric* space with the system of colors that the objection is apparently justified. But there is still a great deal lacking to the establishment of a complete analogy between the space of intuition and the system of color-sensation. Whereas nearly equal distances in sensuous space are immediately recognized as such, a like remark cannot be made of differences of colors, and in this latter province it is not possible to compare physiologically the different portions with one another. And, furthermore, even if there be no difficulty, by resorting to physical experience, in characterizing every color of a system by three numbers, just as the places of geometric space are characterized, and so in creating a metric system similar to the latter, it will nevertheless be difficult to find anything which corresponds to distance or volume and which has an analogous physical significance for the system of colors.

Analogies of Space with Time.

There is always an *arbitrary* element in analogies, for they are concerned with the coincidences to which the attention is directed. But between space and time doubtless the analogy is fully conceded, whether we use the word in its physiological or its physical sense. In both meanings of the term, space is a triple, and time a simple, continuous manifold. A physical event, precisely determined by its conditions, of moderate, not too long or too short duration, seems to us physiologically, *now and at any other time,* as having the same duration. Physical events which at any time are temporarily coincident are likewise temporarily coincident at any other time. Temporal congruence exists, therefore, just as much as does spatial congruence. Unalterable physical temporal objects exist, therefore, as much as unalterable physical spatial objects (rigid bodies). There is not only spatial but there is also temporal substantiality. Galileo employed corporeal phenomena, like the beats of the pulse and breathing, for the determination of time, just as anciently the hands and the feet were employed for the estimation of space.

The simple manifold of *tonal sensations* is likewise analogous to the triple manifold of space-sensations.[1] The comparability of the different parts of the system of tonal sensations is given by the possibility of directly sensing the musical *interval.* A metric system corresponding to geometric space is most easily obtained by expressing tonal pitch in terms of the logarithm of the rate of vibration. For the constant musical interval we have here the expression,

$$\log \frac{n'}{n} = \log n' - \log n = \log \tau - \log \tau' = \text{const.},$$

where n', n denote the rates, and τ', τ the periods of vibration of the higher and the lower note respectively. The difference between the logarithms here represents the constancy of the length on displacement. The unalterable, substantial physical object which we sense as an interval is for the ear *temporally* determined, whereas the analogous object for the senses of sight and touch is spatially determined. Spatial measure seems to us simpler solely because we have chosen for the fundamental measure of geometry distance *itself*, which remains unalterable for sensation, whereas in the province of tones we have reached our measure only by a long and circuitous physical route.

---

[1] My attention was drawn to this analogy in 1863 by my study of the organ of hearing, and I have since then further developed the subject. See my *Analysis of the Sensations.*

Differences of the Analogies.

Having dwelt on the coincidences of our analogized constructs, it now remains for us to emphasize their *differences*. Conceiving time and space as sensational manifolds, the objects whose motions are made perceptible by the alteration of temporal and spatial qualities are characterized by other sensational qualities, as colors, tactual sensations, tones, etc. If the system of tonal sensations is regarded as analogous to the optical space of sense, the curious fact results that in the first province the spatial qualities occur *alone*, unaccompanied by sensational qualities corresponding to the objects, just as if one could see a place or motion without seeing the object which occupied this place or executed this motion. Conceiving spatial qualities as organic sensations which can be excited only *concomitantly* with sensational qualities,[1] the analogy in question does not appear particularly attractive. For the manifold-mathematician, essentially the same case is presented whether an object of definite color moves continuously in optical space, or whether an object spatially fixed passes continuously through the manifold of colors. But for the physiologist and psychologist the two cases are widely different, not only because of what was above adduced, but also, and specifically, because of the fact that the system of spatial qualities is very familiar to us, whereas we can represent to ourselves a system of

---

[1] Compare *supra*, page 14 et seq.

color-sensations only laboriously and artificially, by means of scientific devices. Color appears to us as an excerpted member of a manifold the arrangement of which is in no wise familiar to us.

The Extension of Symbols.

The manifolds here analogized with space are, like the color-system, also threefold, or they represent a *smaller* number of variations. Space contains surfaces as twofold and lines as onefold manifolds, to which the mathematician, generalizing, might also add points as zero-fold manifolds. There is also no difficulty in conceiving analytical mechanics, with Lagrange, as an analytical geometry of four dimensions, time being considered the fourth co-ordinate. In fact, the equations of analytical geometry, in their conformity to the co-ordinates, suggest very clearly to the mathematician the extension of these considerations to an unlimited *larger* number of dimensions. Similarly, physics would be justified in considering an extended material continuum, to each point of which a temperature, a magnetic, electric, and gravitational potential were ascribed, as a portion or section of a multiple manifold. Employment with such symbolic representations must, as the history of science shows us, by no means be regarded as entirely unfruitful. Symbols which initially appear to have no meaning whatever, acquire gradually, after subjection to what might be called intellectual experimenting, a lucid and precise significance. Think only of the negative, fractional, and variable exponents of algebra, or of the cases in which important and vital extensions of ideas have taken place which otherwise would have been totally lost or have made their appearance at a much later date. Think only of the so-called imaginary quantities with which mathematicians long operated, and from which they even obtained important results ere they were in a position to assign to them a perfectly determinate and withal visualizable meaning. But symbolic representation has likewise the disadvantage that the object represented is very easily lost sight of, and that operations are continued with the symbols to which frequently no object whatever corresponds.[1]

[1] As a young student I was always irritated with symbolic deductions of which the meaning was not perfectly clear and palpable. But historical studies are well adapted to eradicating the tendency to mysticism which is so easily fostered and bred by the somnolent employment of these methods, in that they clearly show the heuristic function of them and at the same time elucidate epistemologically the points wherein they furnish their essential assistance. A symbolical representation of a method of calculation has the same significance for a mathematician as a model or a visualisable working hypothesis has for the physicist. The symbol, the model, the hypothesis runs parallel with the thing to be represented. But the parallelism may extend farther, or be extended farther, than was originally intended on the adoption of the symbol. Since the thing represented and the device representing are after all *different*, what would be concealed in the one is apparent in the other. It is scarcely possible to light directly on an operation like $a^{\frac{2}{3}}$. But operating with such symbols leads us to attribute to them an intelligible meaning. Mathematicians worked many years with expressions like $\cos x \times \sqrt{-1} \sin x$ and with exponentials having imaginary exponents before in the struggle for adapting concept and symbol to each other the idea that had been germinating for a century finally found expression in 1806 in Argand, viz., that a relationship could be conceived between magnitude and *direction* by which $\sqrt{-1}$ was represented as a mean direction-proportional between $+1$ and -1.

ANOTHER VIEW OF RIEMANN'S MANIFOLD.

It is easy to rise to Riemann's conception of an n-fold continuous manifold, and it is even possible to realize and visualize portions of such a manifold. Let $a_1, a_2, a_3, a_4 \ldots\ldots a_{n+1}$ be any elements whatsoever (sensational qualities, substances, etc.). If we conceive these elements intermingled in all their possible relations, then each single composite will be represented by the expression

$$\alpha_1 a_1 + \alpha_2 a_2 + \alpha_3 a_3 + \ldots\ldots \alpha_{n+1} a_{n+1} = 1,$$

where the coefficients α satisfy the equation

$$\alpha_1 + \alpha_2 + \alpha_3 + \ldots\ldots \alpha_{n+1} = 1.$$

Inasmuch, therefore, as n of these coefficients α may be selected at pleasure, the totality of the composites of the $n + 1$ elements will represent an n-fold continuous manifold.[1] As co-ordinates of a point of this manifold, we may regard expressions of the form

$$\frac{\alpha_m}{\alpha_1}, \text{ or } f\left(\frac{\alpha_m}{\alpha_1}\right), \text{ for example, } \log\left(\frac{\alpha_m}{\alpha_1}\right).$$

But in choosing definition of distance, or that of any other notion analogous to geometrical concepts, we shall have to proceed very arbitrarily unless *experiences* of the manifold in question inform us that certain metric concepts have a real meaning, and are therefore to be preferred, as is the case for geomet-

[1] If the six fundamental color-sensations were totally independent of one another, the system of color-sensations would represent a five-fold manifold. Since they are contrasted in pairs, the system corresponds to a three-fold manifold.

ric space with the definition[1] derived from the voluminal constancy of bodies for the element of distances $ds^2 = dx^2 + dy^2 + dz^2$, and as is likewise the case for sensations of tone with the logarithmic expression mentioned above. In the majority of cases where such an artificial construction is involved, fixed points of this sort are wanting, and the entire consideration is therefore an ideal one. The analogy with space loses thereby in completeness, fruitfulness, and stimulating power.

Measure of Curvature, and Curvature of Space.

In still another direction Riemann elaborated ideas of Gauss; beginning with the latter's investigations concerning curved surfaces. Gauss's measure of the curvature[2] of a surface at any point is given by the expression $k = \frac{d\sigma}{ds}$ where ds is an element of the surface and $d\sigma$ is the superficial element of the unit-sphere, the limiting radii of which are parallel to the limiting normals of the element ds. This measure of curvature may also be expressed in the form $k = \frac{1}{\rho_1 \rho_2}$, where ρ_1, ρ_2 are the principal radii of curvature of the surface at the point in question. Of special interest are the surfaces whose measure of curvature for all points has the same

[1] Comp. *supra*, p. 73 *et passim.*

[2] *Disquisitiones generales circa superficies curvas*, 1827.

value,—the surfaces of *constant* curvature. Conceiving the surfaces as infinitely thin, non-distensible, but flexible bodies, it will be found that surfaces of like curvature may be made to coincide by bending,—as for example a plane sheet of paper wrapped round a cylinder or cone,—but cannot be made to coincide with the surface of a sphere. During such deformation, nay, even on crumpling, the proportional parts of figures drawn *in the surface* remain invariable as to lengths and angles, provided we do not go out of the two dimensions of the surface in our measurements. Conversely, likewise, the curvature of the surface does not depend on its conformation in the third dimension of space, but solely upon its *interior proportionalities.* Riemann, now, conceived the idea of generalizing the notion of measure of curvature and applying it to spaces of three or more dimensions. Conformably thereto, he assumes that finite unbounded spaces of constant positive curvature are possible, corresponding to the unbounded but finite two-dimensional surface of the sphere, while what we commonly take to be infinite space would correspond to the unlimited plane of curvature zero, and similarly a third species of space would correspond to surfaces of negative curvature. Just as the figures drawn upon a surface of determinate constant curvature can be displaced without distortion upon this surface only (for example, a spherical figure on the surface of its sphere only, or a plane figure in its plane only), so should analogous conditions necessarily hold for spatial figures and rigid bodies. The latter are capable of free motion only in spaces of constant curvature, as Helmholtz[1] has shown at length. Just as the shortest lines of a plane are infinite, but on the surface of a sphere occur as great circles of definite finite length, closed and reverting into themselves, so Riemann conceived in the three-dimensional space of positive curvature analogues of the straight line and the plane as finite but unbounded. But there is a difficulty here. If we possessed the notion of a measure of curvature for a four-dimensional space, the transition to the special case of three-dimensional space could be easily and rationally executed; but the passage from the special to the more general case involves a certain arbitrariness, and, as is natural, different inquirers have adopted here different courses[2] (Riemann and Kronecker). The very fact that for a one-dimensional space (a curved line of any sort) a measure of curvature does not exist having the significance of an interior measure, and that such a measure first occurs in connection with two-dimensional figures, forces upon us the question whether and to what extent something analogous has any meaning for three-dimensional figures. Are we not subject here to an illusion, in that we operate with symbols to which perhaps nothing real corresponds, or at least noth-

[1] "Ueber die Thatsachen, welche der Geometrie zu Grunde liegen." *Göttinger Nachrichten*, 1868, June 3.

[2] Compare, for example, Kronecker, "Ueber Systeme von Functionen mehrerer Variablen." *Ber. d. Berliner Akademie*, 1869.

ing representable to the senses, by means of which we can verify and rectify our ideas?

Thus were reached the highest and most universal notions regarding space and its relations to analogous manifolds which resulted from the conviction of Gauss concerning the empirical foundations of geometry. But the genesis of this conviction has a preliminary history of two thousand years, the chief phenomena of which we can perhaps better survey from the height which we have now gained.

The Early Discoveries in Geometry.

The unsophisticated men, who, rule in hand, acquired our first geometric knowledge, held to the simplest bodily objects (figures): the straight line, the plane, the circle, etc., and investigated, by means of forms which could be conceived as combinations of these simple figures, the connection of their measurements. It could not have escaped them that the mobility of a body is restricted when one and then two of its points are fixed, and that finally it is altogether checked by fixing three of its points. Granting that rotation about an axis (two points), or rotation about a point in a plane, as likewise displacement with constant contact of two points with a straight line and of a third point with a fixed plane laid through that straight line,—granting that these facts were *separately observed*, it would be known how to distinguish between *pure* rotation, *pure* displacement, and the motion compounded of these two independent motions. The first geometry was of course not based on purely metric notions, but made many considerable concessions to the physiological factors of sense.[1] Thus is the appearance explained of two different fundamental measures: the (straight) length and the angle (circular measure). The straight line was conceived as a rigid mobile body (measuring-rod), and the angle as the

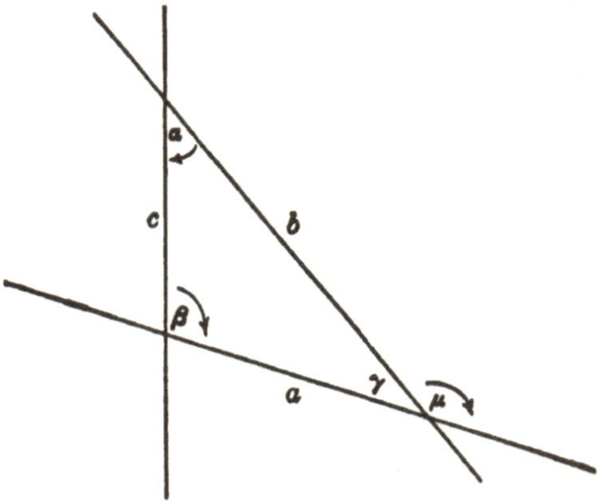

Fig. 14.

rotation of a straight line with respect to another (measured by the arc so described). Doubtless no one ever demanded special proof for the equality of angles at the origin described by the same rotation. Additional propositions concerning angles resulted quite easily. Turning the line b about its intersection with c so as to describe the angle α (Fig. 14), and after coincidence with c turning it again about

[1]Comp. *supra*, p. 83.

its intersection with *a* till it coincides with *a* and so describes the angle β, we shall have rotated *b* from its initial to its final position *a* through the angle μ in the same sense.[1] Therefore the exterior angle $\mu = \alpha + \beta$, and since $\mu + \gamma = 2R$, also $\alpha + \beta + \gamma = 2R$. Displacing (Fig. 15) the rigid system of lines *a*, *b*, *c*, which intersect at 1, within their plane to the position 2, the line *a* always remaining within itself, no alteration of angles will be caused by the mere

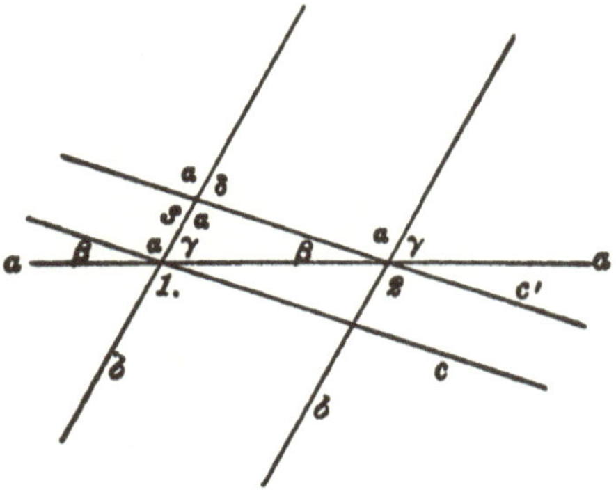

Fig. 15.

motion. The sum of the interior angles of the triangle 1 2 3 so produced is evidently 2 *R*. The same consideration also throws into relief the properties

---

[1] C. R. Kosack, *Beiträge zu einer systematischen Entwickelung der Geometrie aus der Anschauung*, Nordhausen, 1852. I was able to see this programme through the kindness of Prof. F. Pietzker of Nordhausen. Similar simple deductions are found in Bernhard Becker's *Leitfaden für den ersten Unterricht in der Geometrie*, Frankfort on the Main, 1845, and in the same author's treatise *Ueber die Methoden des geometrischen Unterrichts*, Frankfort, 1845. I gained access to the first-named book through the kindness of Dr. M. Schuster of Oldenburg.

of parallel lines. Doubts as to whether successive rotation about several points is equivalent to rotation about *one* point, whether *pure* displacement is at all possible,—which are justified when a surface of curvature differing from zero is substituted for the Euclidean plane,—could never have arisen in the mind of the ingenuous and delighted *discoverer* of these relations, at the period we are considering. The study of the movement of rigid bodies, which Euclid studiously avoids and only covertly introduces in his principle of congruence, is to this day the device best adapted to elementary instruction in geometry. An idea is best made the possession of the learner by the method by which it has been found.

Deductive Geometry.

This sound and naïve conception of things vanished and the treatment of geometry underwent essential modifications when it became the subject of *professional* and *scholarly* contemplation. The object now was to systematize the knowledge of this province for purposes of individual survey, to separate what was directly cognizable from what was deducible and deduced, and to throw into distinct relief the thread of deduction. For the purpose of instruction the simplest principles, those most easily gained and apparently free from doubt and contradiction, are placed at the beginning, and the remainder based upon them. Efforts were made to reduce these initial principles to a minimum, as is observable in the system of Euclid. Through this endeavor to support every notion by another, and to leave to direct knowledge the least possible scope, geometry was gradually detached from the empirical soil out of which it had sprung. People accustomed themselves to regard the derived truths as of higher dignity than the directly perceived truths, and ultimately came to demand proofs for propositions which no one ever seriously doubted. Thus arose,—as tradition would have it, to check the onslaughts of the Sophists,—the system of Euclid with its logical perfection and finish. Yet not only were the ways of research designedly concealed by this artificial method of stringing propositions on an arbitrarily chosen thread of deduction, but the varied organic connection between the principles of geometry was quite lost sight of.[1] This system was more fitted to produce narrow-minded and sterile pedants than fruitful, productive investigators.

[1] Euclid's system fascinated thinkers by its logical excellences, and its drawbacks were overlooked amid this admiration. Great inquirers, even in recent times, have been misled into following Euclid's example in the presentation of the results of their inquiries, and so into actually concealing their methods of investigation, to the great detriment of science. But science is not a feat of legal casuistry. Scientific presentation aims so to expound all the grounds of an idea so that it can at any time be thoroughly examined as to its tenability and power. The learner is not to be led half-blindfolded. There therefore arose in Germany among philosophers and educationists a healthy reaction, which proceeded mainly from Herbart, Schopenhauer, and Trendelenburg. The effort was made to introduce greater perspicuity, more genetic methods, and logically more lucid demonstrations into geometry.

And these conditions were not improved when scholasticism, with its preference for slavish comment on the intellectual products of others, cultivated in thinkers scarcely any sensitiveness for the rationality of their fundamental assumptions and by way of compensation fostered in them an exaggerated respect for the logical form of their deductions. The entire period from Euclid to Gauss suffered more or less from this affection of mind.

Euclid's Fifth Postulate.

Among the propositions on which Euclid based his system is found the so-called Fifth Postulate (also called the Eleventh Axiom and by some the Twelfth): "If a straight line meet two straight lines, so as to make the two interior angles on the same side of it taken together less than two right angles, these straight lines being continually produced, shall at length meet upon that side on which are the angles which are less than two right angles." Euclid easily proves that if a straight line falling on two other straight lines makes the alternate angles equal to each other, the two straight lines will *not* meet but are *parallel*. But for the proof of the converse, that parallels make equal alternate angles with *every* straight line falling on them, he is obliged to resort to the Fifth Postulate. This converse is equivalent to the proposition that *only one* parallel to a straight line can be drawn through a point. Further, by the fact that with the aid of this converse it can be proved that the sum of the angles of a triangle is equal to two right angles and that from this last theorem again the first follows, the relationship between the propositions in question is rendered distinct and the fundamental significance of the Fifth Postulate for Euclidean geometry is made plain.

The intersection of slowly converging lines lies without the province of construction and observation. It is therefore intelligible that in view of the great importance of the assertion contained in the Fifth Postulate the successors of Euclid, habituated by him to rigor, should, even in ancient times, have strained every nerve to demonstrate this postulate, or to replace it by some immediately obvious proposition. Numberless futile efforts were made from Euclid to Gauss, to deduce this Fifth Postulate from the other Euclidean assumptions. It is a sublime spectacle which these men offer: laboring for centuries, from a sheer thirst for scientific elucidation, in quest of the hidden sources of a truth which no person of theory or of practice ever really doubted! With eager curiosity we follow the pertinacious utterances of the ethical power resident in this human search for knowledge, and with gratification we note how the inquirers gradually are led by their failures to the perception that the true basis of geometry is experience. We shall content ourselves with a few examples.

Saccheri's Theory of Parallels.

Among the inquirers notable for their contributions to the theory of parallels are the Italian Saccheri and the German mathematician Lambert. In order to render their mode of attack intelligible, we will remark first that the existence of rectangles and squares, which we fancy we constantly observe, cannot be demonstrated without the aid of the Fifth Postulate. Let us consider, for example, two congruent isosceles triangles *ABC, DBC,* having right angles at *A* and *D* (Fig. 16), and let them be laid together at their hypothenuses *BC* so as to form the

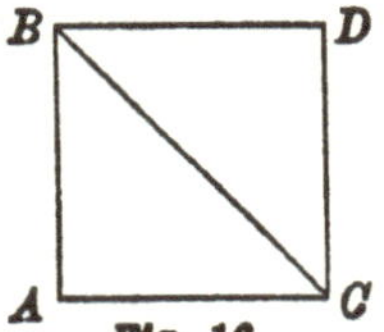

Fig. 16.

equilateral quadrilateral *ABCD*; the first twenty-seven propositions of Euclid do not suffice to determine the character and magnitude of the two equal (right) angles at *B* and *C*. For measure of length and measure of angle are fundamentally different and directly not comparable; hence the first propositions regarding the connection of sides and angles are *qualitative* only, and hence the imperative necessity of a *quantitative* theorem regarding angles, like that of the angle-sum. Be it further remarked that theorems analogous to the twenty-seven planimetric propositions of Euclid may be set up for the surface of a sphere and for surfaces of constant negative curvature, and that in these cases the analogous construction gives respectively obtuse and acute angles at *B* and *C*.

Saccheri's cardinal achievement was his form of stating the problem.[1] If the Fifth Postulate is involved in the remaining assumptions of Euclid, then it will be possible to prove without its aid that in the quadrilateral *ABCD* (Fig. 17) having right angles at *A* and *B* and $AC = BD$, the angles at *C* and *D* likewise are right angles. And, on the other hand, in this event, the assumption that *C* and *D*

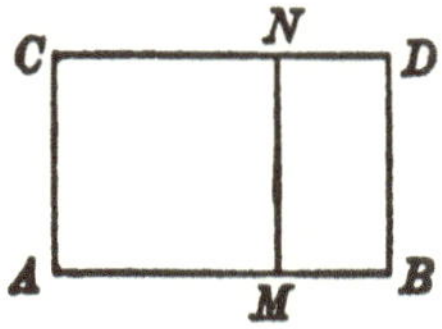

Fig. 17.

are either obtuse or acute will lead to contradictions. Saccheri, in other words, seeks to draw conclusions from the hypothesis of the right, the obtuse, or the acute angle. He shows that each of these hypotheses will hold in all cases if it be proved to hold in one. It is needful to have only one triangle with its angles $\lesseqgtr 2R$ in order to demonstrate the universal validity of the hypothesis of the acute, the right, or the obtuse angle. Notable is the fact that Saccheri also adverts to *physico-geometrical* experi-

[1] *Euclides ab omni naevo vindicatus.* Milan, 1733. German translation in Engel and Staeckel's *Die Theorie der Parallellinien.* Leipsic, 1895.

ments which support the hypothesis of the right angle. If a line CD (Fig. 17) join the two extremities of the equal perpendiculars erected on a straight line AB, and the perpendicular dropped on AB from any point N of the first line, viz., NM, be equal to $CA = DB$, then is the hypothesis of the right angle demonstrated to be correct. Saccheri rightly does not regard it as self-evident that the line which is equidistant from another straight line is itself a straight line. Think only of a circle parallel to a great circle on a sphere which does not represent a

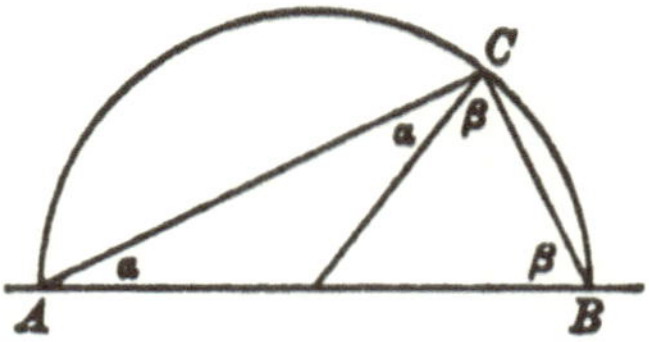

Fig. 18.

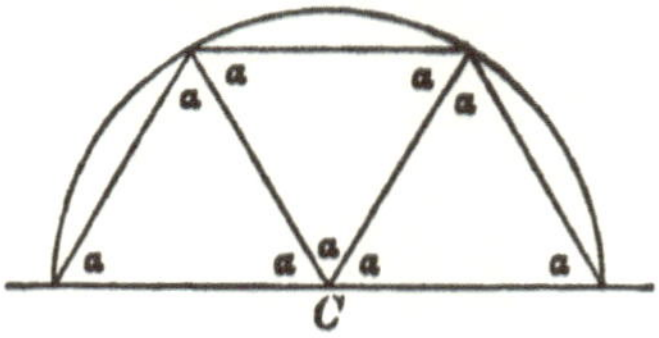

Fig. 19.

shortest line on a sphere and the two faces of which cannot be made congruent.

Other experimental proofs of the correctness of the hypothesis of the right angle are the following. If the angle in a semicircle (Fig. 18) is shown to be a right angle, $\alpha + \beta = R$, then is $2\alpha + 2\beta = 2R$, the sum of the angles of the triangle ABC. If the radius be subtended thrice in a semicircle and the line joining the first and the fourth extremity pass through the center, we shall have at C (Fig. 19) $3\alpha = 2R$, and consequently each of the three triangles will have the angle-sum $2R$. The existence of equiangular triangles of different sizes (similar triangles) is likewise subject to experimental proof. For (Fig. 20) if the angles at B and C give $\beta + \delta + \gamma + \epsilon = 4R$, so also is $4R$ the angle-sum of the quadrilateral $BCB'C'$. Even Wallis[1] (1663) based his proof of the Fifth Postulate on the assumption of the existence of similar triangles, and a modern geometer, Delbœuf, deduced from the assumption of similitude the entire Euclidean geometry.

The hypothesis of the obtuse angle, Saccheri fan-

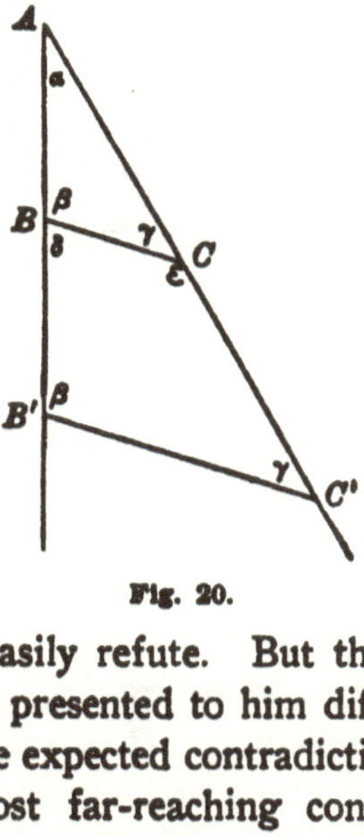

Fig. 20.

cied he could easily refute. But the hypothesis of the acute angle presented to him difficulties, and in his quest for the expected contradictions he was carried to the most far-reaching conclusions, which Lobachévski and Bolyai subsequently rediscovered by methods of their own. Ultimately he felt compelled to reject the last-named hypothesis as incompatible with the nature of the straight line; for it

[1] Engel and Staeckel, *loc. cit.*, p. 21 et seq.

led to the assumption of different kinds of straight lines, which met at infinity, that is, had there a common perpendicular. Saccheri did much in anticipation and promotion of the labors that were subsequently to elucidate these matters, but exhibited withal toward the traditional views a certain bias.

Lambert's Investigations.

Lambert's treatise[1] is allied in method to that of Saccheri, but it proceeds farther in its conclusions, and gives evidence of a less constrained vision. Lambert starts from the consideration of a quadrilateral with three right angles, and examines the consequences that would follow from the assumption that the fourth angle was right, obtuse, or acute. The similarity of figures he finds to be incompatible with the second and third assumptions. The case of the obtuse angle, which requires the sum of the angles of a triangle to exceed $2R$, he discovers to be realized in the *geometry of spherical surfaces,* in which the difficulty of parallel lines entirely vanishes. This leads him to the conjecture that the case of the acute angle, where the sum of the angles of a triangle is less than $2R$, might be realized on the surface of a sphere of imaginary radius. The amount of the departure of the angle-sum from $2R$ is in both cases proportional to the area of the triangle, as may be demonstrated by appropriately di-

[1] Published in 1766. Engel and Staeckel, *loc cit.*, p. 152 et seq.

viding large triangles into small triangles, which on diminution may be made to approach as near as we please to the angle-sum $2R$. Lambert advanced very closely in this conception to the point of view of modern geometers. Admittedly a sphere of imaginary radius, $r\sqrt{-1}$ is not a visualizable geometric construct, but analytically it is a surface having a negative constant Gaussian measure of curvature. It is evident again from this example how experimenting with *symbols* also may direct inquiry to the right path, in periods where other points of support are entirely lacking and where every helpful device must be esteemed at its worth.[1] Even Gauss appears to have thought of a sphere of imaginary radius, as is obvious from his formula for the circumference of a circle (*Letter to Schumacher,* July 12, 1831). Yet in spite of all, Lambert actually fancied he had approached so near to the proof of the Fifth Postulate that what was lacking could be easily supplied.

View of Gauss.

We may turn now to the investigators whose views possess a most radical significance for our conception of geometry, but who announced their opinion only briefly, by word of mouth or letter. "Gauss regarded geometry merely as a logically consistent system of constructs, with the theory of parallels placed at the pinnacle as an axiom; yet he had

[1] See note, p. 104.

reached the conviction that this proposition could not be proved, though it was known from *experience*,—for example, from the angles of the triangle joining the Brocken, Hohenhagen, and Inselsberg, —that it was approximately correct. But if this axiom be not conceded, then, he contends, there results from its non-acceptance a different and entirely independent geometry, which he had once investigated and called by the name of the Anti-Euclidean

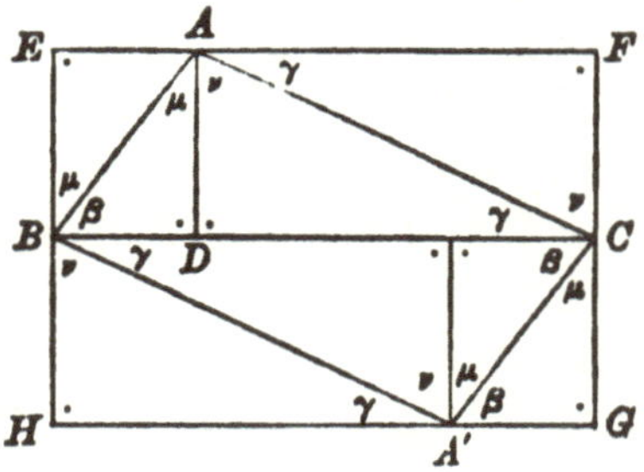

Fig. 21.

geometry." Such, according to Sartorius von Waltershausen, was the view of Gauss.[1]

Researches of Stolz.

Starting at this point, O. Stolz, in a small but very instructive pamphlet,[2] sought to deduce the principal propositions of the Euclidean geometry from the purely observable facts of experience. We shall reproduce here the most important point of Stolz's brochure. Let there be given (Fig. 21) *one*

[1] *Gauss zum Gedächtniss*, Leipsic, 1856.

[2] "Das letzte Axiom der Geometrie," *Berichte des naturw.-medicin. Vereins zu Innsbruck*, 1886, pp. 25-34.

large triangle *ABC* having the angle-sum 2*R*. We draw the perpendicular *AD* on *BC*, complete the figure by *BAE* $\cong$ *ABD* and *CAF* $\cong$ *ACD*, and add to the figure *BCFAE* the congruent figure *CBHA'G*. We obtain thus a *single* rectangle, for the angles *E, F, G, H* are right angles and those at *A, C, A'*, *B* are straight angles (equal to 2*R*), the boundary lines therefore straight lines and the opposite sides equal. A rectangle can be divided into two congruent rectangles by a perpendicular erected at the middle point of one of its sides, and by continuing this procedure the line of division may be brought

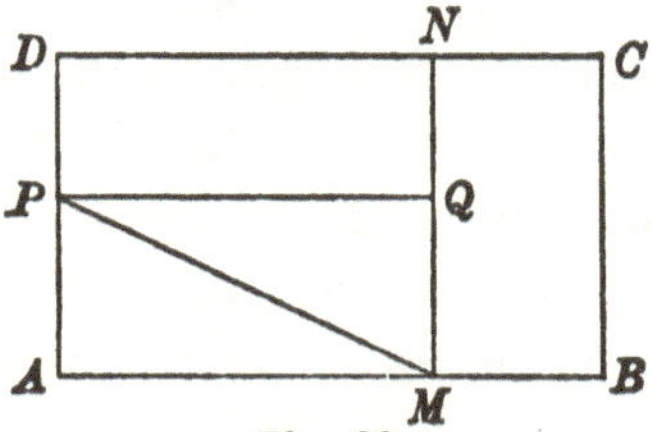

Fig. 22.

to any point we please in the divided side. And the same holds true of the other two opposite sides. It is possible, therefore, from a given rectangle *ABCD* (Fig. 22) to cut out a smaller *AMPQ* having sides bearing any proportion to one another. The diagonal of this last divides it into two congruent *right-angled* triangles, of which each, independently of the ratio of the sides, has the angle-sum 2*R*. Every oblique-angled triangle can by the drawing of a perpendicular be decomposed into right-angled triangles, each of which can again be decomposed into right-angled triangles having smaller sides,—so that $2R$, therefore, results for the angle-sum of *every* triangle if it holds true exactly of *one*. By the aid of these propositions which repose on observation we *conclude* easily that the two opposite sides of a rectangle (or of any so-called parallelogram) are everywhere, no matter how far prolonged, the same distance apart, that is, never intersect. They have the properties of the Euclidean *parallels*, and may be called and *defined* as such. It likewise *follows*, now, from the properties of triangles and rectangles, that two straight lines which are cut by a third straight line so as to make the sum of the interior angles on the same side of them less than two right angles will meet on that side, but in either direction from their point of intersection will move indefinitely far away from each other. The straight line therefore is *infinite*. What was a *groundless* assertion stated as an axiom or an initial principle may as *inference* have a sound meaning.

Geometry and Physics Compared.

Geometry, accordingly, consists of the application of mathematics to experiences concerning space. Like mathematical physics, it can become an exact deductive science only on the condition of its representing the objects of experience by means of schematizing and idealizing concepts. Just as mechanics can assert the constancy of masses or reduce the interactions between bodies to *simple* accelerations *only within the limits of errors of observation,* so likewise the existence of straight lines, planes, the amount of the angle-sum, etc., can be maintained only on a similar restriction. But just as physics sometimes finds itself constrained to replace its ideal assumptions by other more general ones, viz., to put in the place of a constant acceleration of falling bodies one dependent on the distance, instead of a constant quantity of heat a variable quantity,—so a similar procedure is permissible in geometry, when it is demanded by the facts or is necessary temporarily for scientific elucidation. And now the endeavors of Legendre, Lobachévski, and the two Bolyais, the younger of whom was probably indirectly inspired by Gauss, will appear in their right light.

The Contributions of Lobachevski and Bolyai.

Of the labors of Schweickart and Taurinus, also contemporaries of Gauss, we will not speak. Lobachévski's works were the first to become known to the thinking world and so productive of results (1829). Very soon afterward the publication of the younger Bolyai appeared (1833), which agreed in all essential points with Lobachévski's, departing from it only in the form of its developments. According to the originals which have been made almost completely accessible to us in the beautiful editions of Engel and Staeckel,[1] it is permissible to

[1] *Urkunden zur Geschichte der nichteuklidischen Geometrie.* I. N. I. Lobatschefskij. Leipzig, 1899.

assume that Lobachévski also undertook his investigations in the hope of becoming involved in contradictions by the rejection of the Euclidean axiom. But after he found himself mistaken in this expectation, he had the ***intellectual courage*** to draw all the consequences from this fact. Lobachévski gives his conclusions in synthetic form. But we can fairly well imagine the general analyzing considerations that paved the way for the construction of his geometry.

From a point lying outside a straight line *g* (Fig. 23) a perpendicular *p* is dropped and through the

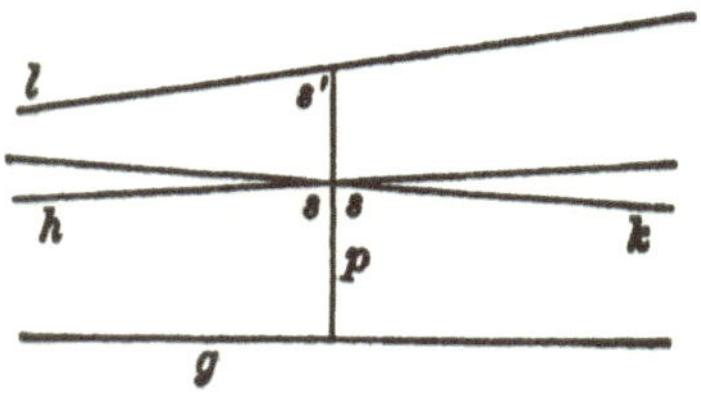

Fig. 23.

same point in the plane *pg* a straight line *h* is drawn, making with the perpendicular an acute angle *s*. Making tentatively the assumption that *g* and *h* do not meet but that on the slightest diminution of the angle *s* they would meet, we are at once forced by the homogeneity of space to the conclusion that a ***second*** line *k* having the same angle *s* similarly deports itself on the other side of the perpendicular. Hence all non-intersecting lines drawn through the same point are situate between *h* and *k*. The latter form the ***boundaries*** between the intersecting and non-intersecting lines and are called by Lobachévski *parallels.*

In the Introduction to his *New Elements of Geometry* (1835) Lobachévski proves himself a thorough natural inquirer. No one would think of attributing even to an ordinary man of sense the crude view that the "parallel-angle" was very much less than a right angle, when on slight prolongation it could be distinctly seen that they would intersect. The relations here considered admit of representation only in drawings that distort the true proportions, and we have rather to picture to ourselves that in the dimensions of the illustration the variation of s from a right angle is so small that h and k are to the eye undistinguishably coincident. Prolonging, now, the perpendicular p to a point beyond its intersection with h, and drawing through its extremity a new line l parallel to h and therefore parallel also to g, it follows that the parallel-angle s' must necessarily be less than s, if h and l are not again to fulfill the conditions of the Euclidean case. Continuing in the same manner, the prolongation of the perpendicular and the drawing of parallels, we obtain a parallel-angle that constantly decreases. Considering, now, parallels which are more remote and consequently converge more rapidly on the side of convergence, we shall logically be compelled to assume, not to be at variance with the preceding supposition, that on approach or on the decrease of the length of the perpendicular the parallel-angle will again increase. The angle of parallelism, therefore, is an inverse function of the perpendicular p, and has been designated by Lobachévski by $\Pi(p)$. A group of parallels in a plane has the arrangement shown schematically in Figure 24. They all approach one another asymptotically toward the side of their convergence. The homogeneity of space requires that every "strip" between two parallels can be made to coincide with every other strip provided it be displaced the requisite distance in a longitudinal direction.

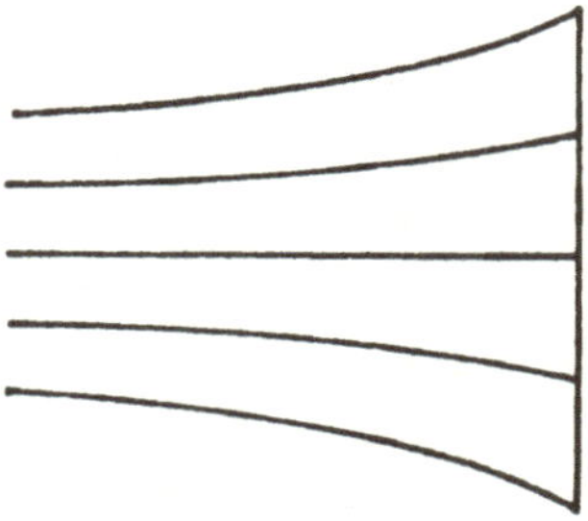

Fig. 24.

If a circle be imagined to increase indefinitely, its radii will cease to intersect the moment the increasing arcs reach the point where the convergence of the radii corresponds to parallelism. The circle then passes over into the so-called "*boundary-line.*" Similarly the surface of a sphere, if it indefinitely increase, will pass into what Lobachévski calls a "*boundary-surface.*" The boundary-lines bear a relation to the boundary-surface analogous to that which a great circle bears to the surface of a sphere. The geometry of the surface of a sphere is independent of the axiom of parallels. But since it can be demonstrated that triangles formed from boundary-lines on a boundary-surface no more exhibit an excess of angle-sum than do finite triangles on a sphere of infinite radius, consequently the rules of the Euclidean geometry likewise hold good for these boundary-triangles. To find points of the boundary-line, we determine (Fig. 25) in a bundle of parallels, $a\alpha$, $b\beta$, $c\gamma$, $d\delta$...... lying in a plane points *a, b, c, d* in each of these parallels so situated with respect to the point *a* in $a\alpha$

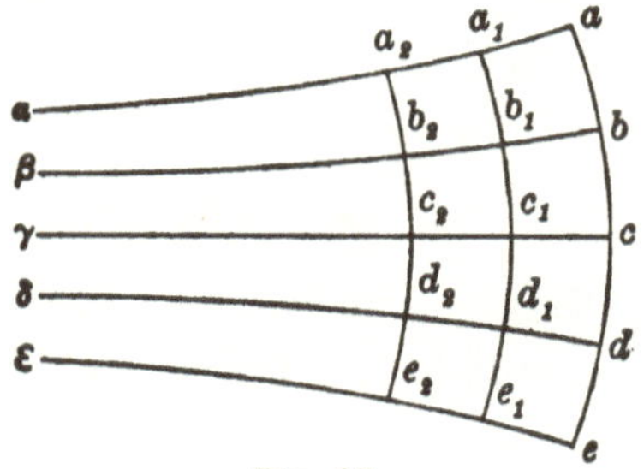

Fig. 25.

that $\angle \alpha ab = \angle \beta ba$, $\angle \alpha ac = \angle \gamma ca$, $\angle \alpha ad = \angle \delta da$.....Owing to the sameness of the entire construction, each of the parallels may be regarded as the "*axis*" of the boundary line, which will generate, when revolved about this axis, the boundary-surface. Likewise each of the parallels may be regarded as the axis of the boundary-surface. For the same reason all boundary-lines and all boundary-surfaces are *congruent*. The intersection of every plane with the boundary-surface is a *circle*; it is a boundary-line only when the cutting plane contains the axis. In the Euclidean geometry there is no boundary-line, nor boundary-surface. The analogues of them are here the straight line and the plane. If no boundary-line exists, then necessarily must any three points not in a straight line lie on a circle. Hence the younger Bolyai was able to replace the Euclidean axiom by this last postulate.

Let $a\alpha$, $b\beta$, $c\gamma$ be a system of parallels, and ae, a_1e_1, a_2e_2..a system of boundary-lines, each of which systems divides the other into equal parts (Fig. 25). The ratio to each other of any two boundary-arcs between the same parallels, e. g., the arcs $ae = u$ and $a_2e_2 = u'$, is dependent therefore solely on their distance apart $aa_2 = x$. We may put generally $\frac{u}{u'} = e^{\frac{x}{k}}$, where k is so chosen that e shall be the base of the Naperian system of logarithms. In this manner exponentials and by means of these hyperbolic functions are introduced. For the angle of parallelism we obtain $s = cot\frac{1}{2}\Pi(p) = e^{\frac{p}{k}}$. If $p = o$, $s = \frac{\pi}{2}$; if $p = \infty$, $s = 0$.

An example will illustrate the relation of the Lobachévskian to the Euclidean and spherical geometries. For a rectilinear Lobachévskian triangle having the sides a, b, c, and the angles A, B, C, we obtain, when C is a right angle,

$$\sinh\frac{a}{k} = \sinh\frac{c}{k}\sin A.$$

Here *sinh* stands for the hyperbolic sine,

$$\sinh x = \frac{e^x - e^{-x}}{2}$$

whereas $$\sin x = \frac{e^{xi} - e^{-xi}}{2i},$$

or, $$\sinh x = \frac{x}{1!} + \frac{x^3}{3!} + \frac{x^5}{5!} + \frac{x^7}{7!} + \ldots\ldots,$$

and $$\sin x = \frac{x}{1!} - \frac{x^3}{3!} + \frac{x^5}{5!} - \frac{x^7}{7!} + \ldots\ldots$$

Considering the relations $\sin(xi) = i$ ($\sinh x$), or $\sinh (xi) = i \sin x$, involved in the foregoing formulæ, it will be seen that the above-given formula for the Lobachévskian triangle passes over into the formula holding for the *spherical* triangle, viz., $\sin \frac{a}{k} = \sin \frac{c}{k} \sin A$, when ki is put in the place of k in the former and k is considered as the radius of the sphere, which in the usual formulæ assumes the value unity. The re-transformation of the spherical formula into the Lobachévskian by the same method is obvious. If k be very great in comparison with a and c, we may restrict ourselves to the first member of the series for sinh or sin, obtaining in both cases, $\frac{a}{k} = \frac{c}{k} \sin A$ or $a = c \sin A$, the formula of *plane Euclidean* geometry, which we may regard as a limiting case of both the Lobachévskian and spherical geometries for very large values of k, or for $k = \infty$. It is likewise permissible to say that all three geometries coincide in the domain of the infinitely small.

---

[1] F. Engel, *N. I. Lobatschefskij, Zwei geometrische Abhandlungen*, Leipsic, 1899.

The Different Systems of Geometry.

As we see, it is possible to construct a self-consistent, non-contradictory system of geometry solely on the assumption of the convergence of parallel lines. True, there is not a single observation of the geometrical facts accessible to us that speaks in favor of this assumption, and admittedly the hypothesis is at so great variance with our geometrical instinct as easily to explain the attitude toward it of the earlier inquirers like Saccheri and Lambert. Our imagination, dominated as it is by our modes of visualizing and by the familiar Euclidean concepts, is competent to grasp only piecemeal and gradually Lobachévski's views. We must suffer ourselves to be led here rather by mathematical *concepts* than by *sensuous images* derived from a single narrow portion of space. But we must grant, nevertheless, that the quantitative mathematical concepts by which we through our own initiative and within a certain arbitrary scope represent the facts of geometrical experience, do not reproduce the latter with absolute exactitude. Different ideas can express the facts with the same exactness in the domain accessible to observation. The *facts* must hence be carefully distinguished from the *intellectual* constructs the formation of which they suggested. The latter—concepts—must be *consistent* with observation, and must in addition be *logically* in accord with one another. Now these two requirements can be fulfilled in *more than one* manner, and hence the different systems of geometry.

Manifestly the labors of Lobachévski were the outcome of intense and protracted mental effort, and it may be surmised that he first gained a clear conception of his system from general considerations and by analytic (algebraic) methods before he was able to present it synthetically. Expositions in this cumbersome Euclidean form are by no means alluring, and it is possibly due mainly to this fact that the significance of Lobachévski's and Bolyai's labors received such tardy recognition.

Lobachévski developed only the consequences of the modification of Euclid's Fifth Postulate. But if we abandon the Euclidean assertion that "two straight lines cannot enclose a space," we shall obtain a companion-piece to the Lobachévskian geometry. Restricted to a surface, it is the geometry of the surface of a sphere. In place of the Euclidean straight lines we have great circles, all of which intersect twice and of which each pair encloses two spherical lunes. There are therefore no parallels. Riemann first intimated the possibility of an analogous geometry for three-dimensional space (of positive curvature),—a conception that does not appear to have occurred even to Gauss, possibly owing to his predilection for infinity. And Helmholtz,[1] who continued the researches of Riemann physically, neglected in his turn, in his first publication, the de-

[1] "Ueber die thatsächlichen Grundlagen der Geometrie," *Wissensch. Abhandl.*, 1866. II., p. 610 et seq.

velopment of the Lobachévskian case of a space of negative curvature (with an imaginary parameter *k*). The consideration of this case is in point of fact more obvious to the mathematician than it is to the physicist. Helmholtz treats in the publication mentioned only the Euclidean case of the curvature zero and Riemann's space of positive curvature.

Applicability of the Different Systems to Reality.

We are able, accordingly, to represent the facts of spatial observation with all possible precision by both the Euclidean geometry and the geometries of Lobachévski and Riemann, provided in the two latter cases we take the parameter k large enough. Physicists have as yet found no reason for departing from the assumption $k = \infty$ of the Euclidean geometry. It has been their practice, the result of long and tried experience, to adhere steadfastly to the *simplest* assumptions until the facts forced their complication or modification. This accords likewise with the attitude of all great mathematicians toward *applied* geometry. The deportment of physicists and mathematicians toward these questions is in the main different, but this is explained by the circumstance that for the former class of inquirers the physical facts are of most significance, geometry being for them merely a convenient implement of investigation, while for the latter class these very questions are the main material of research, and of greatest technical and particularly epistemological interest. Supposing a mathematician to have modified tentatively the simplest and most immediate assumptions of our geometrical experience, and supposing his attempt to have been productive of fresh insight, certainly nothing is more natural than that these researches should be prosecuted farther from a purely mathematical interest. Analogues of the geometry we are familiar with, are constructed on broader and more general assumptions for any number of dimensions, with no pretension of being regarded as more than intellectual scientific experiments and with no idea of being applied to reality. In support of my remark it will be sufficient to advert to the advances made in mathematics by Clifford, Klein, Lie, and others. Seldom have thinkers become so absorbed in revery, or so far estranged from reality, as to imagine for our space a number of dimensions *exceeding the three of the given space of sense,* or to conceive of representing that space by any geometry that departs appreciably from the Euclidean. Gauss, Lobachévski, Bolyai, and Riemann were perfectly clear on this point, and cannot certainly be held responsible for the grotesque fictions which were subsequently constructed in this domain.

It little accords with the principles of a physicist to make suppositions regarding the deportment of geometrical constructs in infinity and in non-accessible places, then subsequently to compare them with our immediate experience and adapt them to it. He prefers, like Stolz, to regard what is directly given as the source of his ideas, which he likewise considers applicable to what is inaccessible until obliged to change them. But he too may be extremely grateful for the discovery that there exist *several* sufficing geometries, that we can make shift also with a *finite* space, etc.,—grateful in short, for the abolition of certain *conventional barriers* of thought.

If we lived on the surface of a planet with a turbid, opaque atmosphere and if, on the supposition that the surface of the earth was a plane and our only instruments were square and chain, we were to undertake geodetic measurements; then the increase in the excess of the angle-sum of large triangles would soon compel us to substitute a spherometry for our planimetry. The *possibility* of analogous experiences in three-dimensional space the physicist cannot as a matter of *principle* reject, although the phenomena that would compel the acceptance of a Lobachévskian or a Riemannian geometry would present so odd a contrast with those to which we have been hitherto accustomed, that no one will regard their actual occurrence as *probable*.

The question whether a given *physical* object is a straight line or the arc of a circle is not properly formulated. A stretched chord or a ray of light is certainly neither the one nor the other. The question is simply whether the object so spatially reacts that it conforms better to the one concept than to the other, and whether with the exactitude which is sufficient for us and obtainable by us it conforms at all to any geometric concept. Excluding the latter case, the question arises, whether we can in practice remove, or at least in thought determine and make allowance for, the *deviation* from the straight line or circle, in other words, *correct* the result of the measurement. But we are dependent always, in practical measurements, on the comparison of *physical* objects. If on direct investigation these coincided with the geometric concepts to the highest attainable point of accuracy, but the indirect results of the measurement deviated more from the theory than the consideration of all possible errors permitted, then certainly we should be obliged to *change* our physico-metric notions. The physicist will do well to await the occurrence of such a situation, while in the meantime the mathematician may be allowed full and free scope for his speculations.

The Concepts of Mathematics and Physics.

Of all the concepts which the natural inquirer employs, the *simplest* are the concepts of space and time. Spatial and temporal objects conforming to his conceptual constructs can be framed with great *exactness*. Nearly every observable *deviation* can be eliminated. We can imagine any spatial or temporal construct realized without doing violence to any fact. The other physical properties of bodies are so intimately interconnected that in their case arbitrary fictions are subjected to narrow restrictions by the facts. A perfect gas, a perfect fluid, a perfectly elastic body does not exist; the physicist knows that his fictions conform only approximately and by arbitrary simplifications to the facts; he is perfectly aware of the deviation, which cannot be removed. We can conceive a sphere, a plane, etc., constructed *with unlimited exactness,* without running counter to any fact. Hence, when any new physical fact occurs which renders a modification of our concepts necessary, the physicist always prefers to sacrifice the less perfect concepts of physics rather than the simpler, more perfect, and more lasting concepts of geometry, which form the solidest foundation of all his theories.

But the physicist can derive in another direction substantial assistance from the labors of geometers. Our geometry refers always to objects of sensuous experience. But the moment we begin to operate with mere things of thought like atoms and molecules, which from their very nature *can never be made the objects of sensuous contemplation,* we are under no obligation whatever to think of them as standing in spatial relationships which are peculiar to the Euclidean three-dimensional space of our sensuous experience. This may be recommended to the special attention of thinkers who deem atomistic speculations indispensable.[1]

---

[1] While still an upholder of the atomic theory, I sought to explain the line-spectra of gases by the vibrations of the atomic constituents of a gas-molecule with respect to another. The difficulties which I here encountered suggested to me (1863) the idea that non-sensuous things did not necessarily have to

The Relativity of All Spatial Relations.

Let us go back in thought to the origin of geometry in the practical needs of life. The recognition of the spatial substantiality and spatial invariability of spatial objects in spite of their movements is a biological necessity for human beings, for spatial quantity is related directly to the quantitative satisfaction of our needs. When knowledge of this sort is not sufficiently provided for by our physiological organization, we employ our hands and feet for comparing the spatial objects. When we begin to compare *bodies* with one another, we enter the domain of physics, whether we employ our hands or an artificial measure. All *physical* determinations are *relative*. Consequently, likewise all *geometrical* determinations possess validity only *relatively* to the measure. The concept of measurement is a concept of relation, which contains nothing not contained in the measure. In geometry we simply assume that the measure will always and everywhere coincide with that with which it has at some other time and in some other place coincided. But this assumption is determinative of nothing con-

be pictured in our sensuous space of three dimensions. In this way I also lighted upon analogues of spaces of different numbers of dimensions. The collateral study of various physiological manifolds (see footnote on page 98 of this book) led me to the problems discussed in the conclusion of this paper. The notion of finite spaces, converging parallels, etc., which can come only from a historical study of geometry, was at that time remote from me. I believe that my critics would have done well had they not overlooked the italicised paragraph. For details see the notes to my *Erhaltung der Arbeit*, Prague, 1872.

cerning the measure. In place of spatial *physiological* equality is substituted an altogether differently defined *physical* equality, which must not be confounded with the former, no more than the indications of a thermometer are to be identified with the sensation of heat. The practical geometer, it is true, determines the dilatation of a heated measure, by means of a measure kept at a constant temperature, and takes account of the fact that the relation of congruence in question is disturbed by this non-spatial physical circumstance. But to the pure theory of space all assumptions regarding the measure are foreign. Simply the physiologically created habit of regarding the measure as invariable is tacitly but unjustifiably retained. It would be quite superfluous and meaningless to assume that the measure, and therefore bodies generally, suffered alterations on displacement in space, or that they remained unchanged on such displacement,—a fact which in its turn could only be determined by the use of a new measure. The *relativity* of all spatial relations is made manifest by these considerations.

Introduction of the Notion of Number.

If the criterion of spatial equality is substantially modified by the introduction of measure, it is subjected to a still further modification, or intensification, by the introduction of the notion of *number* into geometry. There is nicety of distinction gained by this introduction which the idea of congruence alone could never have attained. The application of arithmetic to geometry leads to the notion of *incommensurability* and *irrationality*. Our geometric concepts therefore contain adscititious elements not intrinsic to space; they represent space with a certain latitude, and, arbitrarily also, with greater precision than spatial observation alone could possibly ever realize. This imperfect contact between fact and concept explains the possibility of different systems of geometry.[1]

Significance of the Metageometric Movement.

The entire movement which led to the transformation of our ideas of geometry must be characterized as a sound and healthful one. This movement, which began centuries ago but is now greatly intensified, is not to be looked upon as having terminated. On the contrary, we are quite justified in the expectation that it will long continue, and redound not only to the great advancement of mathematics and geometry, especially in an epistemological regard, but also to that of the other sciences. This movement was, it is true, powerfully stimulated by a few eminent men, but it sprang, nevertheless, not from an individual, but from a general need. This will be seen from the difference in the pro-

[1] It would be too much to expect of matter that it should realize all the atomistic fantasies of the physicist. So, too, space, as an object of experience, can hardly be expected to satisfy all the ideas of the mathematician, though there be no doubt whatever as to the general value of their investigations.

fessions of the men who have taken part in it. Not only the mathematician, but also the philosopher and the educationist, have made considerable contributions to it. So, too, the methods pursued by the different inquirers are not unrelated. Ideas which Leibnitz[1] uttered recur in slightly altered form in Fourier,[2] Lobachévski, Bolyai, and H. Erb.[3] The philosopher Ueberweg,[4] closely approaching in his opposition to Kant the views of the psychologist Beneke,[5] and in his geometrical ideas starting from Erb (which later writer mentions K. A. Erb[6] as his predecessor) anticipates a goodly portion of Helmholtz's labors.

Summary.

The results to which the preceding discussion has led, may be summarized as follows:

1. The source of our geometric concepts has been found to be experience.

2. The character of the concepts satisfying the

[1] See above pp. 66-67.

[2] *Séances de l'Ecole Normale. Débats.* Vol. I., 1800, p. 28.

[3] H. Erb, Grossherzoglich Badischer Finanzrath, *Die Probleme der geraden Linie, des Winkels und der ebenen Fläche,* Heidelberg, 1846.

[4] "Die Principien der Geometrie wissenschaftlich dargestellt." *Archiv für Philologie und Pädagogik.* 1851. Reprinted in Brasch's *Welt- und Lebensanschauung F. Ueberwegs,* Leipzig, 1889, pp. 263-317.

[5] *Logik als Kunstlehre des Denkens,* Berlin, 1842, Vol. II., pp. 51-55.

[6] *Zur Mathematik und Logik,* Heidelberg, 1821. I was unable to examine this work.

same geometrical facts has been shown to be many and varied.

3. By the comparison of space with other manifolds, more general concepts have been reached, of which the geometric represents a special case. Geometric thought has thus been freed from conventional limitations, heretofore imagined insuperable.

4. By the demonstration of the existence of manifolds allied to but different from space, entirely new questions have been suggested. What is space physiologically, physically, geometrically? To what are its specific properties to be attributed, since others are also conceivable? Why is space three-dimensional, etc.?

With questions such as these, though we must not expect the answer to-day or to-morrow, we stand before the entire profundity of the domain to be investigated. We shall say nothing of the inept strictures of the Bœotians, whose coming Gauss predicted, and whose attitude determined him to reserve. But what shall we say to the acrid and captious criticisms to which Gauss, Riemann and their associates have been subjected by men of highest standing in the scientific world? Have these men never experienced in their own persons the truth that inquirers on the outermost boundaries of knowledge frequently discover many things that will not slip smoothly into all heads, but which are not on that account arrant nonsense? True, such inquirers are liable to error, but even the errors of some men are often more fruitful in their consequences than the discoveries of others.

INDEX.